Practice Papers for SQA Exams

Higher

Chemistry

Published by
Leckie & Leckie Ltd
An imprint of HarperCollins*Publishers*
Westerhill Road, Bishopbriggs, Glasgow, G64 2QT
T: 0844 576 8126   F: 0844 576 8131
leckieandleckie@harpercollins.co.uk   www.leckieandleckie.co.uk

A CIP Catalogue record for this book is available from the British Library.

Questions and answers in this book do not emanate from SQA. All of our entirely new and original Practice Papers have been written by experienced authors working directly for the publisher.

Acknowledgement
The following organisation has kindly given permission to use its copyright material free of charge:
Honda (UK) – photo of car in Exam 1, question 13.

MIX
Paper from
responsible sources
FSC **FSC® C007454**
www.fsc.org

FSC™ is a non-profit international organisation established to promote the responsible management of the world's forests. Products carrying the FSC label are independently certified to assure consumers that they come from forests that are managed to meet the social, economic and ecological needs of present and future generations, and other controlled sources.

Find out more about HarperCollins and the environment at
www.harpercollins.co.uk/green

Introduction

Layout of the Book

This book contains practice exam papers, which mirror the actual SQA exam as closely as possible in question style, level and layout. It is the perfect way to familiarise yourself with what the exam papers you will sit will look like.

The answer section at the back of the book contains worked answers to questions, letting you know exactly where marks are gained in an answer and how the right answer is arrived at. It also includes practical tips on how to tackle certain types of questions, details of how marks are awarded and advice on just what the examiners will be looking for.

Revision advice is provided in this introductory section of the book, so please read on!

How To Use This Book

The Practice Papers can be used in two main ways:

1. You can complete an entire practice paper as preparation for the final exam. If you would like to use the book in this way, you might want to complete each practice paper under exam-style conditions by setting yourself a time for each paper and answering it as well as possible without using any references or notes. Alternatively, you can answer the practice paper questions as a revision exercise, using your notes to produce a model answer. Your teacher may mark these for you.

2. You can use the Topic Index at the front of this book to find all the questions within the book that deal with a specific topic. This allows you to focus specifically on areas that you particularly want to revise or, if you are mid-way through your course, it lets you practise answering exam-style questions for just those topics that you have studied.

Revision Advice

Work out a revision timetable for each week's work in advance – remember to cover all of your subjects and to leave time for homework and breaks. For example:

Day	6pm–6.45pm	7pm–8pm	8.15pm–9pm	9.15pm–10pm
Monday	Homework	Homework	English Revision	Chemistry Revision
Tuesday	Maths Revision	Physics Revision	Homework	Free
Wednesday	Geography Revision	Modern Studies Revision	English Revision	French Revision
Thursday	Homework	Maths Revision	Chemistry Revision	Free
Friday	Geography Revision	French Revision	Free	Free
Saturday	Free	Free	Free	Free
Sunday	Modern Studies Revision	Maths Revision	Chemistry Revision	Homework

Make sure that you have at least one evening free a week to relax, socialise and re-charge your batteries. It also gives your brain a chance to process the information that you have been feeding it all week.

Arrange your study time into one-hour or 30-minute sessions, with a break between sessions, e.g. 6pm – 7pm, 7.15pm – 7.45pm, 8pm – 9pm. Try to start studying as early as possible in the evening when your brain is still alert and be aware that the longer you put off starting, the harder it will be to start!

Study a different subject in each session, except for the day before an exam.

Do something different during your breaks between study sessions – have a cup of tea, or listen to some music. Don't let your 15 minutes expand into 20 or 25 minutes though!

Have your class notes and any textbooks available for your revision to hand as well as plenty of blank paper, a pen, etc. You may like to make keyword sheets like the geography example below:

Keyword	Meaning
Anticyclone	An area of high pressure
Secondary Industry	Industries which manufacture things
Erosion	The process of wearing down the landscape

Finally forget or ignore all or some of the advice in this section if you are happy with your present way of studying. Everyone revises differently, so find a way that works for you!

Transfer Your Knowledge

As well as using your class notes and textbooks to revise, these practice papers will also be a useful revision tool as they will help you to get used to answering exam-style questions. As you work through the questions you may find that they refer to a case study or an example that you haven't come across before. Don't worry! You should be able to transfer your knowledge of a topic or theme to a new example. The enhanced answer section at the back will demonstrate how to read and interpret the question to identify the topic being examined and how to apply your course knowledge in order to answer the question successfully.

Command Words

In the practice papers and in the exam itself, a number of command words will be used in the questions. These command words are used to show you how you should answer a question – some words indicate that you should write more than others. If you familiarise yourself with these command words, it will help you to structure your answers more effectively.

Command Word	Meaning/Explanation
Name, state, identify, list	Giving a list is acceptable here – as a general rule you will get one mark for each point you give
Suggest	Give more than a list – perhaps a proposal or an idea
Outline	Give a brief description or overview of what you are talking about
Describe	Give more detail than you would in an outline, and use examples where you can
Explain	Discuss why an action has been taken or an outcome reached – what are the reasons and/or processes behind it.
Justify	Give reasons for your answer, stating why you have taken an action or reached a particular conclusion.
Define	Give the meaning of the term.
Compare	Give the key features of 2 different items or ideas and discuss their similarities and/or their differences.

In the Exam

Watch your time and pace yourself carefully. Work out roughly how much time you can spend on each answer and try to stick to this.

Be clear before the exam what the instructions are likely to be, e.g. how many questions you should answer in each section. The practice papers will help you to become familiar with the exam's instructions.

Read the question thoroughly before you begin to answer it – make sure you know exactly what the question is asking you to do. If the question is in sections, e.g. 15a, 15b, 15c, etc, make sure that you can answer each section before you start writing.

Plan your answer by jotting down keywords, a mindmap or reminders of the important things to include in your answer. Cross them off as you deal with them and check them before you move on to the next question to make sure that you haven't forgotten anything.

Don't repeat yourself as you will not get any more marks for saying the same thing twice. This also applies to annotated diagrams which will not get you any extra marks if the information is repeated in the written part of your answer.

Give proper explanations. A common error is to give descriptions rather than explanations. If you are asked to explain something, you should be giving reasons. Check your answer to an 'explain' question and make sure that you have used plenty of linking words and phrases such as 'because', 'this means that', 'therefore', 'so', 'so that', 'due to', 'since' and 'the reason is'.

Use the resources provided. Some questions will ask you to 'describe and explain' and provide an example or a case study for you to work from. Make sure that you take any relevant data from these resources.

Good luck!

Topic Index

A = Section A (multiple choice) question
B = Section B (extended answer) question

SQA Unit	Topic	Exam 1	Exam 2	Exam 3	Have difficulty	Still needs work	OK	Have difficulty	Still needs work	OK
Unit 1 – Energy Matters	Reaction Rates	A: 6 B: 15b	A: 8 B: 3b, 6c	A: 7 B: 3a, 3b, 4a, 11b						
	Enthalpy	A: 7, 8 B: 16a, 16b	A: 6, 7 B: 5a, 5b	A: 6 B: 3c, 5b						
	Patterns in the Periodic Table	A: 10 B: 10a, 10b,	B: 1a, 1b, 1c	A: 10 B:						
	Bonding, Structure and properties of Elements	A: 11	A: 11 B: 13a	A: 1, 9, 11, 13 B: 1a, 1b						
	Bonding, Structure and properties of Compounds	A: 1, 9, 12, 13 B: 11	A: 2, 9, 10, 12 B: 13b	A: 8 B: 6d						
	The Mole Calculations	A: 14, 15, 16 B: 6a, 6b	A: 13, 14, 15, 16, 17 B: 10c	A: 12, 14, 15 B: 7c, 9b						
	PPA 1 – Reaction Rate (Concentration)			B: 4b, 4c, 4d						
	PPA 2 – Reaction Rate (Temperature)		B: 6a, 6b							
	PPA 3 – Enthalpy of Combustion	B: 16a, 16b		B: 5a						

(continued overleaf)

Unit 2 – The World of Carbon	Fuels	**A:** 17 **B:** 1b, 13b	**A:** 18 **B:** 11c	**A:** 17, 18 **B:** 13c						
	Hydrocarbons	**A:** 19, 22, 23 **B:** 1a, 2b	**A:** 23, 24 **B:** 2a, 2b, 9b, 12a, 12b, 12c	**A:** 19, 21, 24, 30 **B:** 13b, 14a						
	Alcohols, Aldehydes and Ketones	**A:** 20, 21 **B:** 7a, 7d	**A:** 19, 20, 21 **B:**	**A:** 20, 23, 25 **B:** 14c, 14d, 14e						
	Carboxylic Acids and Esters	**A:** 18 **B:** 5c	**A:** 22, 36 **B:** 8d, 8e, 14b	**A:** **B:** 2c						
	Percentage Yield	**B:** 5d	**B:** 8f	**B:**						
	Polymers	**A:** 24, 25 **B:** 2a, 2b	**A:** 25 **B:** 2c	**A:** 26, 27 **B:** 12b						
	Natural Products	**A:** 26, 27, 28	**A:** 26, 27, 28 **B:** 9a, 9c, 9d, 14a, 14b	**A:** 28, 29 **B:** 2a, 2b, 2c, 11b, 11c						
	PPA 1 – Aldehydes and Ketones	**B:** 7b, 7c		**B:** 14d, 14e						
	PPA 2 – Esters	**B:** 5a, 5b		**B:**						
	PPA 3 – Decomposition of Hydrogen peroxide	**B:** 8b		**B:** 11a						
Unit 3 – Chemical Reactions	The Chemical Industry	**A:** 29	**A:** 29	**A:** 31						
	Hess's Law	**A:** 30 **B:** 4c	**A:** 30 **B:** 12d	**A:** 32 **B:** 13d						
	Equilibrium	**A:** 31, 32 **B:** 15c	**A:** 31, 32	**A:** 33, 34						
	Acids and Bases	**A:** 33, 34, 35 **B:** 12a	**A:** 22, 33, 34, 35, 36 **B:** 8b, 8c	**A:** 4, 35, 36, 37 **B:** 10						
	Redox Reactions	**A:** 36, 37 **B:** 13a, 14bi, 14bii	**A:** 37, 38 **B:** 11a, 15a, 15b	**A:** 38 **B:** 6a, 12c						
	Electrolysis	**B:** 13c	**B:** 4a, 4b	**B:** 9b						
	Nuclear Chemistry	**A:** 38, 39, 40 **B:** 3a, 3b, 3c	**A:** 39, 40 **B:**	**A:** 39, 40 **B:** 6c						
	PPA 1 – Hess's Law	**A:** 30								
	PPA 2 – Electrolysis		**B:** 4a, 4b							
	PPA 3 – Redox Titrations	**B:** 14bi, 14bii	**B:** 15a, 15b	**B:**						
	Standard Grade Revision	**A:** 2, 4 **B:** 4a, 12c, 14, 15	**A:** 1, 5 **B:** 3a, 8a, 10a, 11c	**A:** 2, 3, 16, 22 **B:** 7a, 9a						
	Problem Solving	**A:** 3, 5 **B:** 4b, 6c, 8a, 9a, 9b, 12b, 17a, 17b	**A:** 2, 4 **B:** 7a, 7b, 7c, 10b, 11b	**A:** 5 **B:** 4c, 6b, 7b, 7d, 8a, 8d, 12a, 13a, 14b						

Exam 1

Chemistry   Higher

Practice Papers   **Exam 1**
For SQA Exams   **Higher Level**

Fill in these boxes:

Name of centre   Town

Forename(s)   Surname

You may use the Chemistry Higher and Advanced Higher Data Booklet.

SECTION A

Answers should be marked in HB pencil on a copy of the answer sheet on page 46 of this book.

SECTION B

1. Attempt all questions
2. Write you answers clearly, in ink
3. Write rough work in this book, then score it out clearly when you have written the fair copy.
4. The space available for answers does not indicate how much you should write. You do not have to fill all the available space.

Scotland's leading educational publishers

Section A
Attempt all questions in this section.
Mark your answers on a copy of the answer sheet on page 46 of this book.

1. Which of the following substances is a non-conductor of electricity when solid but becomes a good conductor when molten?

    A   Sodium
    B   Sodium chloride
    C   Carbon
    D   Carbon tetrachloride

2. The experiment shown below was performed to study the corrosion of an iron nail.

    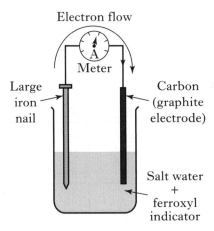

    After a short period of time a blue colour would be visible at

    A   the iron nail.
    B   the carbon electrode.
    C   neither the iron nail or the carbon electrode.
    D   both the iron nail and the carbon electrode.

3. In which of the following compounds do **both** ions have the same electron arrangement as neon?

    A   Lithium oxide
    B   Magnesium chloride
    C   Sodium fluoride
    D   Potassium fluoride

4. What volume of sodium hydroxide solution, concentration of 0.6 mol l^{-1}, is required to neutralise 100 cm^3 of hydrochloric acid, concentration of 0.3 mol l^{-1}?

    A   25 cm^3
    B   50 cm^3
    C   250 cm^3
    D   5 litres

5. Naturally occurring hydrogen consists of two isotopes ^1H and ^2H.
    How many different types of hydrogen molecules can exist?

    A   1
    B   2
    C   3
    D   4

6. The graph below shows the change in the concentration of a reactant with time as the reaction proceeds.

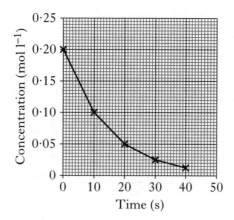

    What is the average rate of this reaction, in mol l^{-1} s^{-1}, for the first 20 seconds?

    A   0·01
    B $2·5 \times 10^{-3}$
    C $7·5 \times 10^{-3}$
    D 5×10^{-3}

7. When 10 g of propan-1-ol (relative formula mass = 60) was burned, 310 kJ of energy was released. Using this information what is the enthalpy of combustion of propan-1-ol in kJ mol^{-1}?

    A   −155
    B   −1550
    C   −1860
    D   −3100

8.   Which row of the following table correctly represents the graph shown?

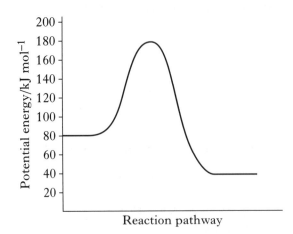

	Activation energy (kJ mol^{-1})	Enthalpy Change
A	140	Exothermic
B	140	Endothermic
C	100	Exothermic
D	100	Endothermic

9.   Which of the following compounds has the **most** ionic character?

A   Magnesium fluoride
B   Sodium fluoride
C   Calcium fluoride
D   Potassium fluoride

10. Which equation represents the first ionisation of magnesium?

A $Mg(s) \longrightarrow Mg^+(s) + e^-$
B $Mg(s) \longrightarrow Mg^+(g) + e^-$
C $Mg(g) \longrightarrow Mg^+(g) + e^-$
D $Mg(g) \longrightarrow Mg^{2+}(g) + 2e^-$

11. Which of the following substances does **not** have a covalent network structure?

A   Diamond
B   Sulphur
C   Silicon dioxide
D   Silicon carbide

12. In which of the following compounds is hydrogen bonding most likely to occur?

    A   Propyl propanoate
    B   Propanoic acid
    C   Propane
    D   Propene

13. Which of the following substances is most likely to be soluble in tetrachloromethane (carbon tetrachloride), CCl_4?

    A   Phosphorus chloride
    B   Calcium chloride
    C   Sodium chloride
    D   Caesium chloride

14. How many moles of water can theoretically be produced from the complete combustion of 2 moles of methane gas?

    A   1
    B   2
    C   3
    D   4

15. Which of the following gases contains the greatest number of molecules?

    A   25 g fluorine
    B   25 g chlorine
    C   25 g oxygen
    D   25 g hydrogen

16. How many ions are present in 3.1 g of sodium oxide?

    A $6 \cdot 02 \times 10^{23}$
    B $9 \cdot 03 \times 10^{23}$
    C $3 \cdot 10 \times 10^{23}$
    D $1 \cdot 806 \times 10^{24}$

17. Which of the following equations shows a reaction that takes place during reforming?

    A $C_2H_5OH \longrightarrow C_2H_4 + H_2O$
    B $C_6H_{14} \longrightarrow C_6H_6 + 4H_2$
    C $C_{10}H_{22} \longrightarrow C_6H_{14} + C_4H_8$
    D $C_4H_8 + H_2 \longrightarrow C_4H_{10}$

18. Shown below is the structure of aspirin.

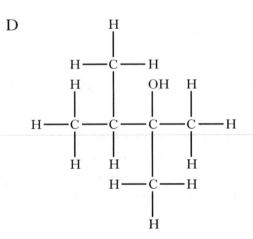

What two functional groups are highlighted?

A   Carboxyl and ester
B   Hydroxyl and carbonyl
C   Hydroxyl and carboxyl
D   Ester and carbonyl

19. Which of the following compounds **does not** have isomeric forms?

A CH_3OH
B $C_2H_4Cl_2$
C C_3H_7Cl
D $C_3H_6Cl_2$

20. Which of the following alcohols is a tertiary alcohol?

21. Compound **X** is oxidised using the oxidising agent, acidified potassium dichromate, to produce a compound with the formula $CH_3CH_2CH_2COOH$

    Which of the following could be compound **X**?

    A   Butan-2-ol
    B   Butan-1-ol
    C   Propan-1-ol
    D   Propan-2-ol

22. Which statement about benzene is correct?

    A   Benzene is an isomer of cyclohexane.
    B   Benzene decolourises bromine solution.
    C   Benzene undergoes addition reactions more readily than ethene.
    D   The carbon to hydrogen ratio in benzene is the same as in ethyne.

23. Syngas is more commonly known as synthesis gas. What is synthesis gas a mixture of?

    A   Carbon monoxide and hydrogen
    B   Carbon dioxide and hydrogen
    C   Carbon monoxide and oxygen
    D   Methane and oxygen

24. Detergents are now available in water-soluble plastic capsules for use in dishwashers and washing machines. Which plastic could be used to make these capsules?

    A   Kevlar
    B   Biopol
    C   Poly(ethenol)
    D   Poly(ethyne)

25. Polymerisation is the process of reacting monomer molecules together to form polymer chains. Which of the following reactants could be added to stop a polymer chain from becoming too long during condensation polymerisation?

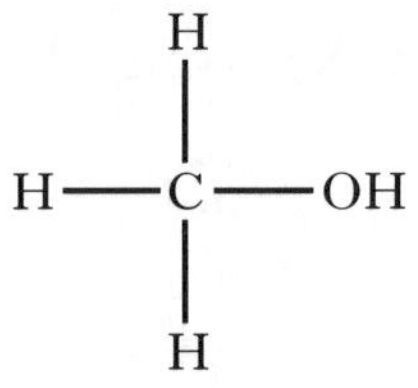

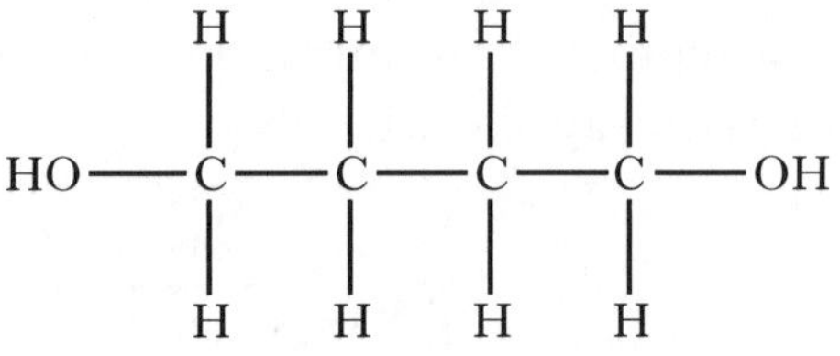

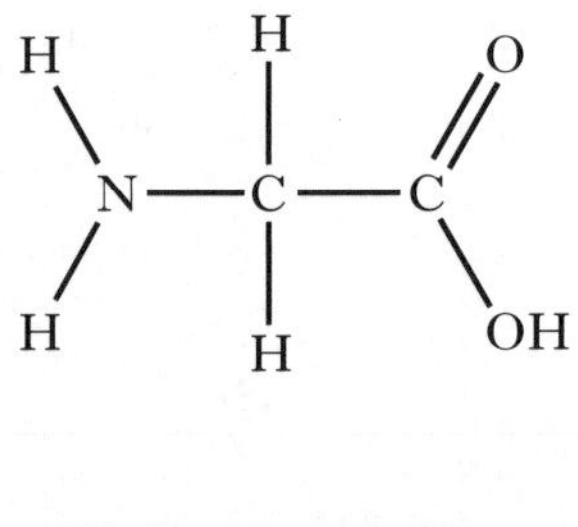

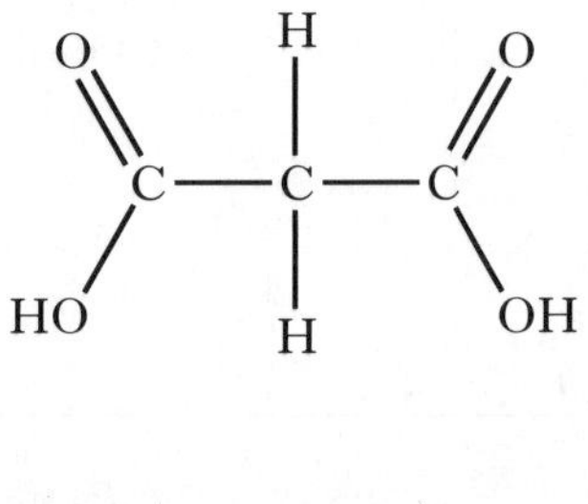

26. Glyceryl trioleate is a typical ester found in oils. It has the formula

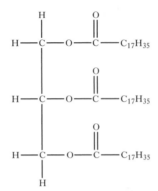

Hydrolysis of this compound produces 3 molecules of oleic acid and 1 molecule of

A   Glycerol
B   Propan-2-ol
C   Water
D   Stearic acid

27. Margarines are hardened fats produced from vegetable oils. The melting point of the oil is increased by

A   hydration
B   hydrogenation
C   hydrolysis
D   dehydration

28. Enzyme molecules can be classified as

A   Alcohols
B   Proteins
C   Esters
D   Fats

29. Which of the following compounds is a raw material in the chemical industry?

A   Iron
B   Ammonia
C   Methane
D   Benzene

30. State the enthalpy of formation of carbon monoxide (reaction X) using the reaction pathway shown.

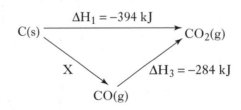

A -110 kJ mol^{-1}
B $+110 \text{ kJ mol}^{-1}$
C -678 kJ mol^{-1}
D $+678 \text{ kJ mol}^{-1}$

31. A catalyst is added to a chemical reaction at equilibrium. Which statement about catalysts is correct?

    A   The ΔH of the reverse reaction will increase.

    B   The ΔH of the forward reaction will increase.

    C   The position of equilibrium is unchanged.

    D   The position of equilibrium shifts to the right.

32. Iodine monochloride and chlorine gas react together to establish the following equilibrium in a closed system.

$$ICl(l) + Cl_2(g) \rightleftharpoons ICl_3(s) \quad \Delta H = -106 \text{ kJ mol}^{-1}$$

Which line in the table shows the conditions that would cause the greatest increase in the amount of ICl_3 produced in this reaction?

	Pressure	Temperature
A	High	High
B	Low	Low
C	High	Low
D	Low	High

33. A 0·01 mol l^{-1} solution of nitric acid can be described as

    A   a concentrated solution of a strong acid

    B   a concentrated solution of a weak acid

    C   a dilute solution of a strong acid

    D   a dilute solution of a weak acid

34. The concentration of OH$^-$ ions in a solution of sodium hydroxide is found to be 0·01 mol l^{-1}. What is the pH of the solution?

    A   5

    B   9

    C   12

    D   14

35. 50 cm^3 of an alkali is required to neutralise 25 cm^3 of 2 mol l^{-1} hydrochloric acid. What volume of the same alkali would be required to neutralise 25 cm^3 of 2 mol l^{-1} ethanoic acid?

    A   25 cm^3

    B   50 cm^3

    C   75 cm^3

    D   100 cm^3

36. During a redox process, chlorate ions are converted into chlorine.

$$ClO_3^- \longrightarrow Cl_2$$

The reaction is carried out in acidic conditions to provide H^+ ions. How many H^+ ions would be required to balance this ion-electron equation?

A   12
B   10
C   8
D   6

37. In which of the following reactions is hydrogen acting as an oxidising agent?

A $C_4H_8 + 5H_2 \longrightarrow C_4H_{10}$
B $2K + H_2 \longrightarrow 2KH$
C $S + H_2 \longrightarrow H_2S$
D $C_4H_6 + 4H_2 \longrightarrow C_4H_8$

38. Radioactive iodine-131 is a very effective treatment of cancer of the thyroid gland. Iodine-131 differs from the stable isotopes of iodine in

A   atomic number
B   atomic mass
C   chemical properties
D   valency

39. 5 g of radioactive sodium-24 is placed into a solution of sodium chloride and the solution soon becomes radioactive.
The radioactivity of the solution is compared to the radioactivity of sodium-24 sample. The radioactivity of the solution was found to be

A   different intensity and the same half-life.
B   different intensity and a different half-life
C   the same intensity and the same half-life
D   the same intensity but a different half-life

40. The equation

$$^{2}_{1}H + ^{3}_{1}H \longrightarrow ^{4}_{2}He + ^{1}_{0}n$$

represents which of the following?

A   nuclear fusion
B   nuclear fission
C   radioactive decay
D   neutron capture

SECTION B

Write your answers clearly in ink.

1. Branched chain hydrocarbons are added to petrol to increase its octane number. An example of one of the branched chain hydrocarbons is shown below

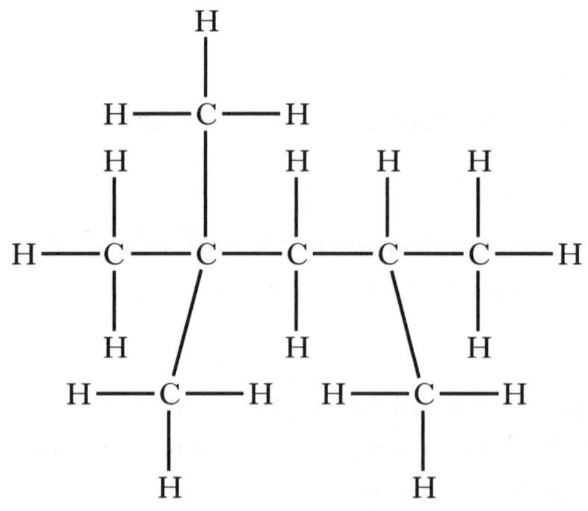

(a) The hydrocarbon shown is known in industry as isooctane. Give the systematic name for this hydrocarbon.   (1 mark)

(b) Give another **type** of hydrocarbon that could be added to petrol to increase its octane number.   (1 mark)

2. Kevlar was a polymer that was developed in 1965 and was first used in the tyres of racing cars to give them strength. It is also used as body armour.

(a) Apart from its strength, give another property of Kevlar that makes it suitable for use as body armour.   (1 mark)

(b) Shown below are the structures of the monomers that are combined to form Kevlar. What name is given to the highlighted functional group?   (1 mark)

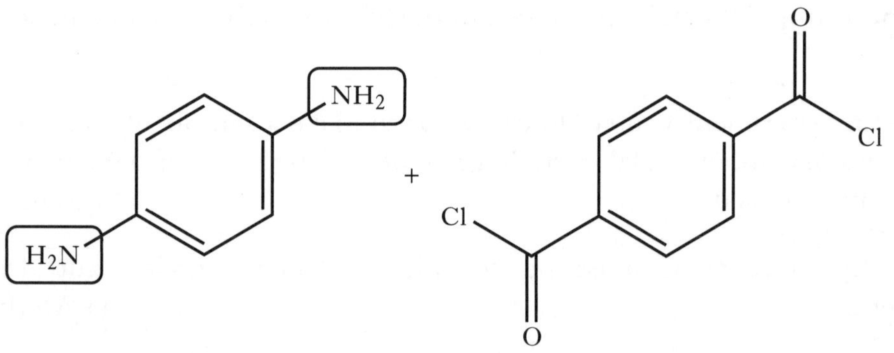

(c) What type of polymerisation reaction forms Kevlar?   (1 mark)

3. When Dmitri Mendeleev developed the periodic table, he left gaps in the table for elements that had not yet been discovered. One of the gaps was filled by the radioactive element Technetium. An isotope of this element Technetium-99 has a very long half-life and is produced as radioactive waste in nuclear reactors. To reduce this danger, it is converted to an isotope with a lower half-life by the equation shown below

$$^{99}_{43}Tc + ^{1}_{0}n \longrightarrow X$$

(a) Identify isotope X (1 mark)

(b) Isotope X decays by beta emission. Write a balanced equation for this reaction. (1 mark)

(c) The new isotope formed has a half-life of 16 s. If 100 g of the sample is left for 48 s, what mass of the sample would remain? (1 mark)

4. Dinitrogen tetroxide is one of the most important rocket propellants ever developed and, although it was developed in the 1950s, it is still in use today. In rockets it is combined with methylhydrazine.

$$5N_2O_4 + 4CH_3NHNH_2 \longrightarrow XCO_2 + YN_2 + ZH_2O$$

(a) Draw the full structural formula of methyhydrazine. (1 mark)

(b) Balance the above equation by giving the values of X, Y and Z. (1 mark)

(c) The equation for the combustion of methylhydrazine is

$$CH_3NHNH_2(l) + 2\frac{1}{2}O_2(g) \longrightarrow CO_2(g) + N_2(g) + 3H_2O(l) \ \Delta H = -1305kJ \ mol^{-1}$$

Using this information and the enthalpies of combustion given in the Data booklet calculate the enthalpy change for the following reaction. (2 Marks)

$$C(s) + N_2(g) + 3H_2(g) \longrightarrow CH_3NHNH_2(l) \ \Delta H=?$$

5. Ethyl ethanoate is an ester compound that has a characteristic sweet smell which is similar to pear drops. It can be produced in the lab from ethanol and ethanoic acid

(a) Describe an experiment you could carry out to make a sample of this ester in the laboratory. Include a labelled diagram of the apparatus and the names of any chemicals used. (3 Marks)

(b) What safety precautions must be considered whilst carrying out this experiment? (1 Mark)

(c) Draw the full structural formula of ethyl ethanoate. (1 Mark)

(d) A student performed this experiment in a lab using 9 g of ethanol and an excess of ethanoic acid. He produced 11 g of the ester. What was the % yield? (3 Marks)

6. John was investigating the use of metals to produce hydrogen as a fuel for cars. He added 10·36 g of lead to 50 cm³ of 1 mol l⁻¹ hydrochloric acid in a boiling tube. The balanced equation for the reaction is:

$$Pb(s) + 2HCl(aq) \longrightarrow PbCl_2(aq) + H_2(g)$$

(a) Show by calculation which reactant is in excess.   (2 Marks)

(b) Calculate the mass of hydrogen gas which will have been given off during the experiment.   (1 Mark)

(c) The hydrogen produced in the reaction can be contaminated with hydrogen chloride vapour, which is very soluble in water. Complete the following diagram to show how the hydrogen chloride could be removed before the hydrogen is collected.   (1 Mark)

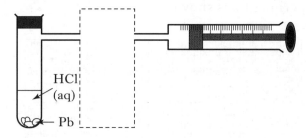

7. Although aldehydes and ketones contain the carbonyl functional group, they have different structures.

(a) In what way is the structure of an aldehyde different from a ketone?   (1 Mark)

(b) Due to their difference in structure only aldehydes react with Benedict's reagent. What colour change would be observed in the reaction?   (1 Mark)

(c) Name another suitable oxidising agent that could be used in this reaction in place of the Benedict's reagent.   (1 Mark)

(d) When butanal is oxidised the compound with the formula C_3H_7COOH is produced. Name this compound.   (1 Mark)

8. A solution of hydrogen peroxide can be used to treat contact lenses. The hydrogen peroxide is then broken down by adding the enzyme catalase. The equation for this reaction is

$$2H_2O_2 \longrightarrow H_2O + O_2$$

This reaction was performed in the lab in three different ways. The results are shown in the table below

Experiment	Volume of H_2O_2	Concentration of H_2O_2 (mol l^{-1})	Catalyst Used
1	25 cm^3	0·4	No
2	25 cm^3	0·4	Yes
3	25 cm^3	0·6	Yes

The results graph of experiment 2 is shown

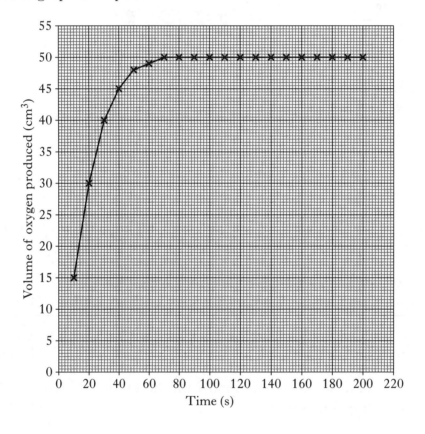

(a) Add curves to the graph to illustrate the results that would be expected in experiments 1 and 3 (2 Marks)

(b) Draw a labelled diagram of the assembled apparatus that could be used to perform this experiment in the lab. (1 Mark)

9. X-ray diffraction is a technique that is used to determine the structures of molecules. They are based on observing the scattering intensity of an X-ray beam hitting a sample and this allows us to determine the shape of a molecule. An example is shown below:

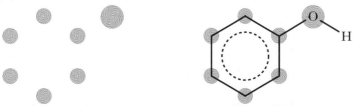

X-ray electron density map   Electron density map over molecule

Hydroxybenzene (phenol), C_6H_5OH

(a) The hydrogen atoms do not show up clearly on an electron density map. Suggest a reason for this. (1 Mark)

(b) Shown below is an electron density map of another molecule.

Draw the full structural formula of this molecule (1 Mark)

10. Magnesium and sulphur are in the same period of the periodic table but the Mg^{2+} ion is much smaller than a S^{2-} ion.

(a) Give the reason for this. (1 Mark)

(b) The sulphur ion S^{2-} has the same electron arrangement as the calcium ion Ca^{2+}. Give the reason why the Ca^{2+} ion is smaller than the sulphur ion. (1 Mark)

11. The table shown compares the properties of ethanol and propane

Hydrocarbon	Boiling Point	State at Room Temperature	Solubility in Water	Formula Mass
Ethanol	78·4°C	Liquid	Soluble	46
Propane	−42°C	Gas	Insoluble	44

Explain **fully** why although propane and ethanol have very similar formula masses, they have very different boiling points and solubility in water. You should give mention to both the bonding and intermolecular forces involved and how these arise. (4 Marks)

12. Sodium bicarbonate has many uses from cooking to medical uses. It is produced in industry by the Solvay process which is the reaction of calcium carbonate, sodium chloride, ammonia and carbon dioxide. The initial reaction in this process is the neutralisation reaction between carbon dioxide and sodium hydroxide to produce the salt sodium carbonate

$$CO_2 + 2NaOH \longrightarrow Na_2CO_3 + H_2O$$

(a) Sodium carbonate is an alkaline salt. What can be concluded about the strengths of the acid and alkali used in the reaction? (1 Mark)

(b) The sodium chloride used in the Solvay process is obtained from seawater and because of this it can be contaminated with magnesium ions. The sodium carbonate produced can be used to remove the magnesium. Why is sodium carbonate solution suitable for the removal of magnesium ions? (1 Mark)

(c) The ammonia used in this process is produced by the reaction of calcium hydroxide and ammonia chloride. This reaction also produces a salt and water. Write a balanced equation for this reaction. (1 Mark)

13. Car companies are looking to fuel cells to power the cars of the future. For example, the sports car shown below has a hydrogen fuel cell rather than a conventional petrol engine.

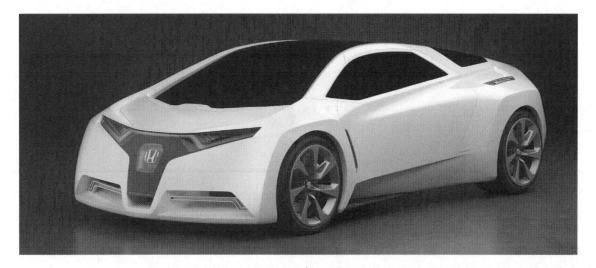

Initially car companies looked at methanol fuel cells but they found some disadvantages with these cells when compared to hydrogen fuel cells.

(a) The ion electron equations for the oxidation and reduction reactions that take place in a methanol fuel cell are:

$$CH_3OH(l) + H_2O(l) \longrightarrow CO_2(g) + 6H^+(aq) + 6e^-$$
$$3O_2(g) + 12H^+(aq) + 12e^- \longrightarrow 6H_2O(l)$$

Combine the two ion-electron equations to give the equation for the overall redox reaction. (1 Mark)

(b) The equation for the redox reaction in a hydrogen fuel cell is

$$2H_2(g) + O_2(g) \longrightarrow 2H_2O(l)$$

Give the disadvantage of using a methanol fuel cell when compared to a hydrogen fuel cell. (1 Mark)

(c) The hydrogen gas that is used in these fuel cells can be produced by the electrolysis of water. The hydrogen gas is produced at the negative electrode

$$H_2O(l) + 2e^- \longrightarrow H_2(g) + 2OH^-(aq)$$

Calculate the volume of hydrogen gas produced when a steady current of 0·5 A is passed through water for 30 minutes. (3 Marks)
(Take the molar volume of hydrogen to be 24 litres)

14. Potassium permanganate has many uses both in and out of the lab. It can be heated to produce oxygen gas. The equation for this reaction is shown

$$KMnO_4(s) \longrightarrow K_2O(s) + MnO_2(s) + O_2(g)$$

(a) Balance the above equation (1 Mark)

(b) Acidified potassium permanganate can be used to determine the concentration of iron(II) sulphate solution by titration.

(i) Suggest two points of good practice that a chemist should follow to ensure that an accurate end-point is obtained in this titration. (1 Mark)

(ii) The results obtained showed that an average of 8·35 cm^3 of iron(II) sulphate solution was required to completely react with 25 cm^3 of 0·2 mol l^{-1} potassium permanganate solution. The equation for this reaction is

$$5Fe^{2+}(aq) + MnO_4^-(aq) + 8H^+(aq) \longrightarrow 5Fe^{3+}(aq) + Mn^{2+}(aq) + 4 H_2O(l)$$

Calculate the concentration of the iron(II) sulphate solution in mol l^{-1}. (2 Marks)

15. Nitrogen dioxide can be produced in a number of ways but industrially it is produced using the Ostwald process. In the first stage of this process nitrogen monoxide is produced by passing ammonia over a hot catalyst.

(a) What **type** of element is used as the catalyst in this process? (1 Mark)

(b) The catalyst used is a heterogeneous catalyst. What is a heterogeneous catalyst? (1 Mark)

(c) In the second stage of this process the nitrogen monoxide combines with oxygen in an exothermic reaction to produce nitrogen dioxide.

$$2NO(g) + O_2(g) \rightleftharpoons 2NO_2(g)$$

What happens to the yield of nitrogen dioxide if the mixture is cooled? (1 Mark)

16. The enthalpy of combustion of ethanol can be calculated using measurements obtained using the apparatus shown.

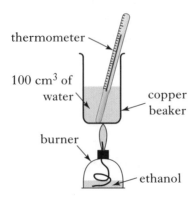

(a) In the experiment it was found that burning 0·64 g of ethanol increased the temperature of the water by 10·0°C. Use these results to calculate the enthalpy of combustion of ethanol. (3 Marks)

(b) Give **two** possible reasons why the result obtained is different to the value of the enthalpy of combustion stated in the data book. (1 Marks)

17. The Born-Haber cycle is an approach to analysing the energies involved in chemical reactions. Born-Haber cycles are used primarily as a means of calculating lattice enthalpies, which cannot otherwise be measured directly. An example is shown below

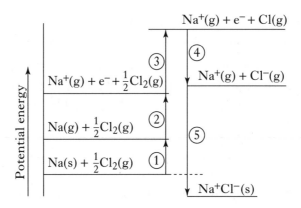

(a) Indicate which of the five reaction stages involves the greatest release of energy. (1 Mark)

(b) Write an equation for the overall reaction that is represented in the Born-Haber cycle. (1 Mark)

Exam 2

Chemistry   Higher

Practice Papers   **Exam 2**
For SQA Exams   **Higher Level**

Fill in these boxes:

Name of centre   Town

Forename(s)   Surname

You may use the Chemistry Higher and Advanced Higher Data Booklet.

SECTION A

Answers should be marked in HB pencil on a copy of the answer sheet on page 46 of this book.

SECTION B

1. Attempt all questions
2. Write you answers clearly, in ink
3. Write rough work in this book, then score it out clearly when you have written the fair copy.
4. The space available for answers does not indicate how much you should write. You do not have to fill all the available space.

Leckie⚡Leckie
Scotland's leading educational publishers

Section A
Attempt all questions in this section.
Mark your answers on a copy of the answer sheet on page 46 of this book.

1. Which of the following compounds contains both a halide ion and a transition metal ion?

   A Sodium hydroxide
   B Magnesium bromide
   C Copper chloride
   D Potassium permanganate

2. Which of the following gases does not react with water to form ions?

   A Ammonia
   B Sulphur dioxide
   C Methane
   D Carbon dioxide

3. Particles with the same electron arrangement are called isoelectronic. Which of the following compounds contain ions that are isoelectronic?

   A Li_2O
   B Na_2S
   C MgF_2
   D $CaBr_2$

4. A mixture of sodium bromide and sodium carbonate is known to contain 0·6 moles of bromide ions and 0·4 moles of carbonate ions. How many moles of sodium ions are present?

   A 0·5
   B 0·8
   C 1·0
   D 1·4

5. A gas is bubbled through a solution of sulphuric acid and the pH increased. The gas could have been

   A Sulphur dioxide
   B Hydrogen
   C Propane
   D Ammonia

6. The potential energy diagram shown below refers to a reversible reaction.

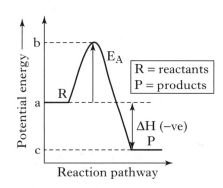

What is the activation energy E_A expressed in terms of a, b, and c for the **reverse** reaction?

A $E_A = b - a$
B $E_A = b + a$
C $E_A = b - c$
D $E_A = b + c$

7. The enthalpy of combustion is always the energy released when

A   1 mole of a fuel is completely burned in excess oxygen
B   An excess of fuel is burned in 1 mole of oxygen
C   1 g of a fuel is completely burned in excess oxygen
D   An excess of fuel is burned in 1 g of oxygen

8. A small increase in the temperature of a reaction mixture results in a large increase in reaction rate. The main reason for this is

A   the activation energy is lowered
B   the enthalpy change is increased
C   the enthalpy change is decreased
D   the kinetic energy of the particles is increased

9. Which of the following compounds exists as discrete molecules?

A   Sulphur dioxide
B   Sodium chloride
C   Silicon dioxide
D   Silver(I) oxide

10. Which of the following chlorides has the **most** ionic character?

A   NaCl
B   CsCl
C $CaCl_2$
D $AlCl_3$

11. An unknown element is found to have a melting point of over 2000°C but an oxide of this element is a gas at room temperature. What type of bonding will be present in the element?

   A   Metallic
   B   Pure covalent
   C   Ionic
   D   Polar covalent

12. Which of the following compounds has polar molecules?

   A   Methane
   B   Ammonia
   C   Carbon dioxide
   D   Carbon tetrachloride

13. Which of the following contains the same number of atoms as 80 g of argon?

   A   32 g of oxygen
   B   4 g of hydrogen
   C   34 g of ammonia
   D   14 g of nitrogen

14. How many moles of ions are present in 2 moles of sodium sulphide?

   A   1
   B   2
   C   3
   D   6

15. Avogadro's constant is equal to the number of

   A   atoms in 131·3 g of xenon
   B   molecules in 32 g of methane
   C   ions in 58·5 g of sodium chloride
   D   formula units in 49·5 g of copper(II) chloride

16. How many ions are present in 47·65 g of magnesium chloride?

   A $6·02 \times 10^{23}$
   B $9·03 \times 10^{23}$
   C $3·01 \times 10^{23}$
   D $1·806 \times 10^{24}$

17. The equation for the complete combustion of methane is shown below.

$$CH_4(g) + 2O_2(g) \longrightarrow CO_2(g) + 2H_2O(l)$$

If 20 cm³ of methane is ignited and burned with 100 cm³ of oxygen, what is the volume of the resulting gas mixture?

   A   20 cm³
   B   50 cm³
   C   80 cm³
   D   120 cm³

18. The petrol fraction obtained from the distillation of crude oil is not ready for use as a fuel but requires further processing to make it burn more efficiently. Which of the following does not increase the octane number of petrol?

A   aromatic hydrocarbons
B   cycloalkanes
C   straight-chain alkanes
D   branched alkanes

19. Which of the following compounds is a ketone?

A

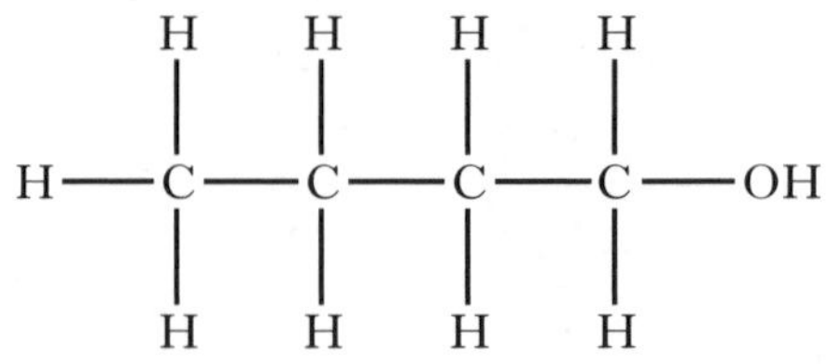

$$H-\overset{\overset{\displaystyle H}{|}}{\underset{\underset{\displaystyle H}{|}}{C}}-\overset{\overset{\displaystyle H}{|}}{\underset{\underset{\displaystyle H}{|}}{C}}-\overset{\overset{\displaystyle H}{|}}{\underset{\underset{\displaystyle H}{|}}{C}}-\overset{\overset{\displaystyle H}{|}}{\underset{\underset{\displaystyle H}{|}}{C}}-OH$$

B

$$H-\overset{\overset{\displaystyle H}{|}}{\underset{\underset{\displaystyle H}{|}}{C}}-\overset{\overset{\displaystyle H}{|}}{\underset{\underset{\displaystyle H}{|}}{C}}-\overset{\overset{\displaystyle H}{|}}{\underset{\underset{\displaystyle H}{|}}{C}}-\overset{O}{\underset{OH}{C}}$$

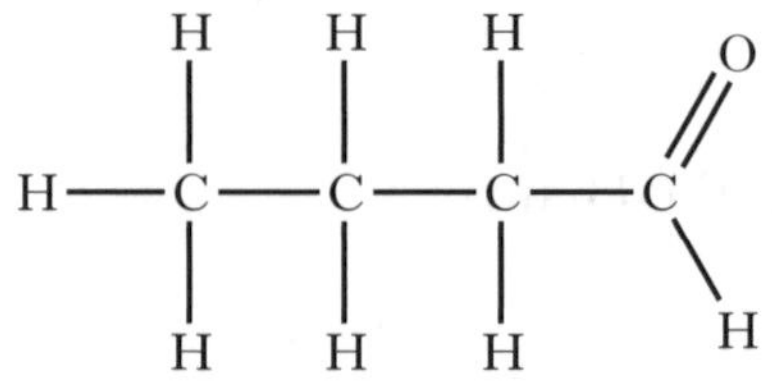

C

$$H-\overset{\overset{\displaystyle H}{|}}{\underset{\underset{\displaystyle H}{|}}{C}}-\overset{\overset{\displaystyle H}{|}}{\underset{\underset{\displaystyle H}{|}}{C}}-\overset{\overset{\displaystyle H}{|}}{\underset{\underset{\displaystyle H}{|}}{C}}-\overset{O}{\underset{H}{C}}$$

D

$$H-\overset{\overset{\displaystyle H}{|}}{\underset{\underset{\displaystyle H}{|}}{C}}-\overset{\overset{\displaystyle H}{|}}{\underset{\underset{\displaystyle H}{|}}{C}}-\overset{\overset{\displaystyle O}{||}}{C}-\overset{\overset{\displaystyle H}{|}}{\underset{\underset{\displaystyle H}{|}}{C}}-H$$

20. Dehydration of hexan-3-ol can produce two isomeric products, hex-2-ene and hex-3-ene. Which of the following alcohols can also produce two isomeric products on dehydration?

A   Propan-2-ol
B   Butan-2-ol
C   Pentan-3-ol
D   Heptan-4-ol

21. Which of the following reactions can be classed as reduction?

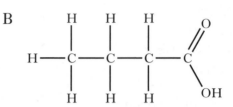
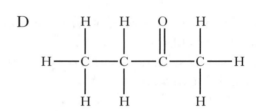

A $CH_3CH_2CH_2CH_2CHO$ $\longrightarrow$ $CH_3CH_2CH_2CH_2COOH$
B $CH_3CH_2CH_2COCH_3$ $\longrightarrow$ $CH_3CH_2CH_2CH(OH)CH_3$
C $CH_3CH_2CH_2CH(OH)CH_3$ $\longrightarrow$ $CH_3CH_2CH_2COCH_3$
D $CH_3CH_2CH(OH)CH_3$ $\longrightarrow$ $CH_3CH_2COCH_3$

22. Which of the following acids would react with potassium hydroxide to produce an alkaline salt?

A H_2SO_4
B HCl
C CH_3COOH
D HNO_3

23. What type of reaction is performed to turn ethene into 1,2-dichloroethane?

    A hydration
    B addition
    C oxidation
    D hydrolysis

24. The use of chlorofluorocarbons has been heavily regulated since the 1970s because of their destructive effect on which of the following?

    A wildlife
    B ozone layer
    C global warming
    D acid rain.

25. Cured polyamide resins

    A are formed by addition polymerisation
    B are used as fibres
    C have a 3-D structure with cross linking between the chains
    D are soluble in water.

26. Glycerol is a chemical compound also commonly called glycerin or glycerine. What is the correct systematic name for glycerol?

    A Propan-2-ol
    B Propane-1,2,3-triol
    C Propanoic acid
    D Propyl propanoate

27. Which of the following types of bond are broken during the hydrolysis of proteins

    A $C=O$
    B $C-N$
    C $C-H$
    D $N-H$

28. Fats have a higher melting point than oils because compared to oils, fats

    A have a higher degree of unsaturation
    B have more loosely packed molecules
    C have less carbon atoms
    D have a higher degree of saturation.

29. Which of the following costs would be classed as a variable cost in the chemical industry?

    A the cost of labour
    B the cost of plant construction
    C the cost of raw materials
    D the cost of land rental.

30. Given that

$$S(s) + O_2(g) \longrightarrow SO_2(g) \qquad \Delta H = -297 \text{ kJ mol}^{-1}$$
$$H_2(g) + \tfrac{1}{2}O_2(g) \longrightarrow H_2O(l) \qquad \Delta H = -286 \text{ kJ mol}^{-1}$$
$$H_2S(g) + 1\tfrac{1}{2}O_2(g) \longrightarrow H_2O(l) + SO_2(g) \qquad \Delta H = -563 \text{ kJ mol}^{-1}$$

The enthalpy change for the reaction

$$H_2(g) + S(s) \longrightarrow H_2S(g)$$

will be

A $-1146 \text{ kJ mol}^{-1}$
B -20 kJ mol^{-1}
C $+20 \text{ kJ mol}^{-1}$
D $+1146 \text{ kJ mol}^{-1}$

31. In a reversible reaction equilibrium is established when

A the concentration of the reactants and products are equal
B the concentration of the reactants and products are constant
C the concentration of only the products remains unchanged
D the concentration of only the reactants remains unchanged.

32. In which of the following reactions would a change in pressure **not** affect the yield of product?

A $ICl(l) + Cl_2(g) \rightleftharpoons ICl_3(s)$
B $2NO_2(g) \rightleftharpoons N_2O_4(g)$
C $H_2(g) + I_2(g) \rightleftharpoons 2HI(g)$
D $N_2(g) + 3H_2(g) \rightleftharpoons 2NH_3(g)$

33. Oxalic acid can be extracted from rhubarb. A solution of this acid is found to have a pH of 4. What is the hydrogen ion concentration, in mol l^{-1}, of this solution?

A 0·01
B 0·001
C 0·0001
D 0·00001

34. Lemon juice has a pH of 2 and orange juice has a pH of 3. The concentrations of H$^+$ ions in lemon juice and orange juice are in the ratio

A 10:1
B 1:10
C 100:1
D 2:3

35. Which line in the table is correct for 1 mol l⁻¹ hydrochloric acid in comparison to 1 mol l⁻¹ ethanoic acid?

	pH	Conductivity
A	Lower	Higher
B	Lower	Lower
C	Higher	Lower
D	Higher	Higher

36. On the structure shown below which hydrogen atom will ionise most easily in aqueous solution?

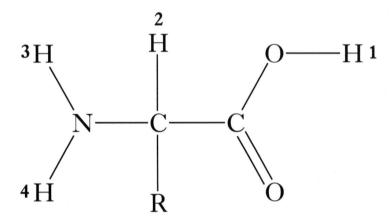

A 1
B 2
C 3
D 4

37. Stannic chloride ($SnCl_4$) was used in World War 1 as a chemical weapon due to its corrosive and toxic properties. It can be produced by the reaction shown

$$HgCl_2 + SnCl_2 \longrightarrow Hg + SnCl_4$$

What ion is oxidised in the above reaction?

A Sn^{4+}
B Sn^{2+}
C Hg^{2+}
D Cl^-

38. Which of the following reactions can be classed as a redox reaction?

A Displacement
B Combustion
C Precipitation
D Fermentation

39. Polonium is an element with the symbol Po and an atomic number of 84. A radioactive isotope of this element is polonium-210. What is the neutron to proton ratio of this isotope?

A   0·4

B   0·67

C   1·5

D   2·5

40. There are three types of radioactive decay all of which can be stopped in different ways as shown in the diagram

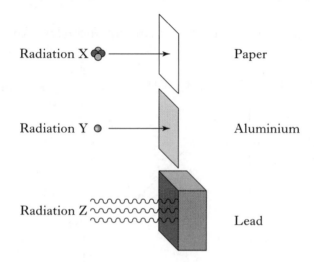

Which line in the table correctly identifies the types of radiation identified by particles X, Y and Z?

	Particle X	Particle Y	Particle Z
A	Alpha	Beta	Gamma
B	Beta	Alpha	Gamma
C	Beta	Gamma	Alpha
D	Gamma	Beta	Alpha

SECTION B

Write your answers clearly in ink.

1. (a) Atoms of different elements have different electronegativities. What is meant by the term electronegativity? (1 Mark)

   (b) Atoms of different elements are different sizes. What is the trend on atomic size going across the periodic table from potassium to krypton? (1 Mark)

   (c) The first ionisation enthalpy of sodium is less than the first ionisation enthalpy of lithium. Explain fully why this is the case. (2 Marks)

2. Reforming is a process carried out in the petroleum industry. An example of reforming is shown.

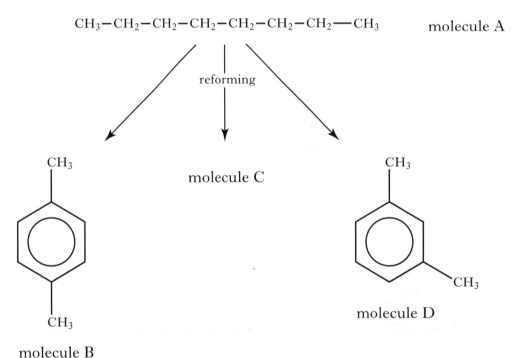

$CH_3-CH_2-CH_2-CH_2-CH_2-CH_2-CH_2-CH_3$ molecule A

reforming

molecule C

molecule B

molecule D

   (a) Name molecule A (1 Mark)

   (b) Molecule C is a branched hydrocarbon. Draw a possible structure for molecule C (1 Mark)

   (c) The structure of poly(vinylcarbazole) is shown below.

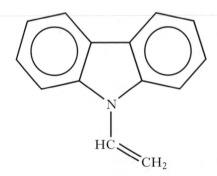

What unusual property does poly(vinyl carbazole) have? (1 Mark)

3. Catalytic converters in car exhausts convert harmful gases into less harmful gases. Two of the less harmful gases formed are a result of a reaction between carbon monoxide and nitrogen monoxide.

    (a) Name the two gases produced as a result of this reaction.   (1 Mark)

    (b) Catalytic converters are an example of heterogeneous catalysis. Explain clearly what happens to the molecules in the exhaust gas during their conversion to less harmful gases.   (2 Marks)

4. Sulphuric acid can be electrolysed using the apparatus shown below.

    Hoffman Apparatus

    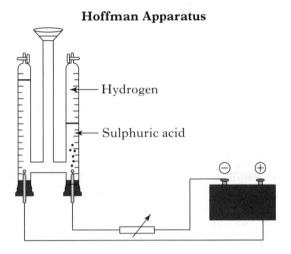

    (a) What three measurements are required to accurately calculate the quantity of electricity required to produce 1 mol of hydrogen gas?   (1 Mark)

    (b) During electrolysis of the acid 0·25 g of hydrogen was collected. What mass of aluminium would be obtained if the same current was used for the same length of time in electrolysis of aluminium chloride?   (3 Marks)

5. Many chemical reactions involve energy changes. These energy changes can have many practical uses.

    (a) Hand warmers make use of an increase in temperature that takes place during the catalytic oxidation of iron in air. Name the type of reaction that results in an increase in temperature?   (1 Mark)

    (b) Flameless ration heaters are used by the US Military to prepare meals when on missions. The energy released is required to heat 227 g of water by 37·8°C in twelve minutes. Calculate the energy required to do this.   (1 Mark)

6. The PPA on the effect of changing the temperature on the rate of a reaction can be studied using the following reaction

    $$5(COOH)_2(aq) + 6H^+(aq) + 2MnO_4^-(aq) \rightarrow 2Mn^{2+}(aq) + 10CO_2(g) + 8H_2O(\ell)$$

    (a) What colour change indicates that the reaction is over?   (1 Mark)

(b) The reaction is carried out at temperatures ranging from about 40°C to 80°C. Why is the reaction not performed at temperatures below and above this range? (1 Mark)

(c) The graph shown illustrates the kinetic energy for molecules at a certain temperature.

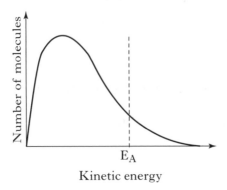

Kinetic energy

Draw a new line on the graph that you would expect to see if the temperature was increased by only 10°C. (1 Mark)

7. Magnesium metal can be extracted from sea water using electrolysis. Seawater contains about 0·13% of magnesium. The flow diagram for this procedure is shown below

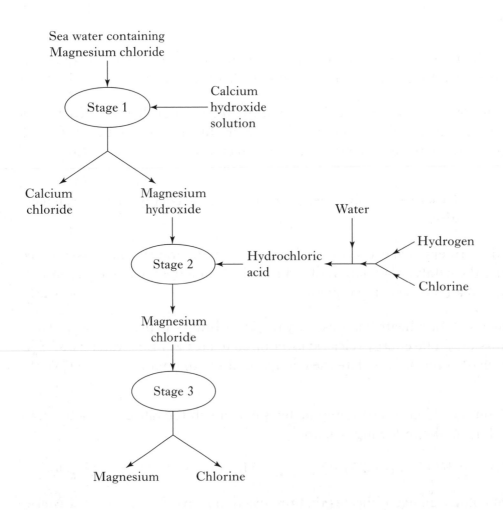

(a) Stage 1 and 2 involve chemical reactions. What name can be given to the reactions at stage 1 and 2? (2 Marks)

(b) What process is taking place at stage 3 to separate the magnesium chloride into its constituent elements. (1 Mark)

(c) Draw an arrow onto the diagram that would make the process more economical. (1 Mark)

8. Fruit preserves contain both fruit juice and pieces of fruit. The ingredients label of a fruit preserve is shown below

    Fruit Preserve – Fruit
    Citric Acid
    Sweeteners (Glucose and sucrose)
    Acidity regulator (sodium citrate)
    Water

(a) What reagent could be used to distinguish between glucose and sucrose? (1 Mark)

(b) Citric acid is classified as a weak acid. What is meant by a weak acid? (1 Mark)

(c) Sodium citrate is a salt of citric acid. Estimate the pH of a solution of sodium citrate. (1 Mark)

(d) The fruit used contains an ester that provides the preserve with flavour. The structure of the ester is shown below.

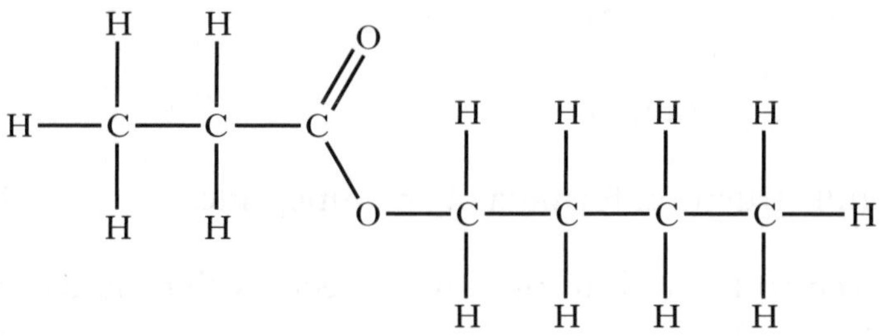

    Name the acid used to produce this ester. (1 Mark)

(e) Name the catalyst used to produce an ester in the lab. (1 Mark)

(f) A group of pupils produced this ester in the lab and managed to obtain a 65% yield. Calculate the mass of ester they produced if they used 22·2 g of alcohol. (Mass of one mole of the alcohol = 74 g; mass of one mole of ester = 130 g) (2 Marks)

9. Phenylalanine is an essential amino acid which is found naturally in the breast milk of mammals.

(a) What is meant by an essential amino acid? (1 Mark)

(b) How many hydrogen atoms does a molecule of phenylalanine contain? (1 Mark)

(c) Alanine can combine with phenylalanine to form a dipeptide.

Alanine

Draw a possible full structural formula of the dipeptide. (1 Mark)

(d) What name is given to the link in protein molecules that holds the amino acid molecules together? (1 Mark)

10. A student carried out two experiments which involved reacting excess marble chips (calcium carbonate) with different concentrations of hydrochloric acid.

(a) Write a balanced equation for the above reaction. (1 Mark)

(b) The first experiment was the reaction of excess marble with 100cm³ of 0.5 mol l⁻¹ hydrochloric acid. The curve obtained for experiment 1 is shown.

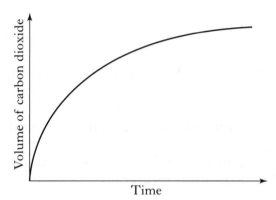

Draw a curve on the graph to show the result you would expect for experiment 2 when 50 cm³ of 1 mol l⁻¹ hydrochloric acid is reacted with the same mass of marble. (1 Mark)

(c) Calculate the mass in grams, of carbon dioxide produced in experiment 2. (2 Marks)

11. Carbon monoxide is a tasteless, colourless but a highly toxic gas. Despite its toxicity, it was once used as a domestic fuel. One method used to produce carbon monoxide is shown by the equation

$$Zn + CaCO_3 \rightarrow ZnO + CaO + CO$$

(a) Write the ion-electron half-equation for the oxidation step in the reaction. (1 Mark)

(b) Carbon monoxide can be produced in the lab as a product of the reaction of carbon dioxide with hot carbon. The carbon dioxide is made by the reaction of dilute hydrochloric acid with solid calcium carbonate. Unreacted carbon dioxide is removed before the carbon monoxide is collected by displacement of water.

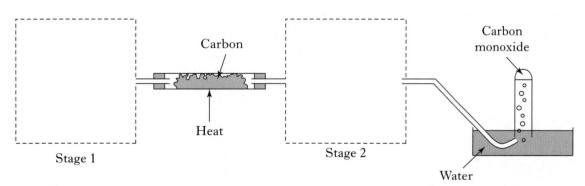

Complete the diagram showing how the carbon dioxide is produced in stage 1 and how the excess carbon dioxide is removed using lime water in stage 2. (2 Marks)

(c) Carbon monoxide can also be found in the exhaust fumes of cars. Why is carbon monoxide produced by car engines? (1 Mark)

12. Alkynes are traditionally known as acetylenes. The first member of the alkynes is ethyne which has the formula C_2H_2.

(a) Due to their functional group alkynes can undergo addition reactions. One example is shown.

$$H—C\equiv C—H \xrightarrow{\text{HCl}} \text{Compound X} \xrightarrow{\text{HCl}} \text{Compound Y}$$

Draw the possible full structural formulas of compounds X and Y (2 Marks)

(b) Draw the full structural formula of one isomer of compound Y (1 Mark)

(c) Vinyl alcohol or ethanol is produced from ethyne by an addition reaction with water. Give another name for this reaction. (1 Mark)

(d) The equation for the enthalpy of formation of ethyne is

$$2C(s) + H_2(g) \longrightarrow C_2H_2(g)$$

Use the enthalpies of combustion of ethyne, carbon and hydrogen given in the Data Booklet to calculate the enthalpy of formation of ethyne. (2 Marks)

13. The properties of four different substances are shown in the table

Substance	Melting Point (°C)	Boiling Point (°C)	Conduction
A	−77	−33	No
B	1883	2503	No
C	773	1407	When molten
D	1538	2862	Yes

(a) Complete the table below using the letters to show the type of bonding present in each substance. (2 Marks)

Substance	Bonding and structure
	Metallic
	Ionic
	Covalent network
	Covalent molecular

(b) Methane has a boiling point of −164°C However, methanol has a boiling point of 65°C. Explain clearly why methanol has a relatively high boiling point compared to methane.   (2 Marks)

14. Hydrogenation results in the conversion of liquid vegetable oils to solid or semi-solid fats, such as those present in margarine.

(a) What is meant by the term hydrogenation?   (1 Mark)

(b) What is the main class of organic compound found in fats and oils?   (1 Mark)

(c) The best-selling margarine contains 0·1 g of sodium per 10 g of margarine. Sodium is present in the margarine as sodium chloride. Calculate the mass of salt a 500 g tub of margarine contains.   (2 Marks)

15. Hydrogen peroxide is used medically as an antiseptic for cleaning wounds. The concentration of a hydrogen peroxide solution can be calculated by titration with acidified potassium permanganate. The reaction is represented by the equation

$$2MnO_4^-(aq) + 6H^+(aq) + 5H_2O_2(aq) \longrightarrow 2Mn^{2+}(aq) + 8H_2O(l) + 5O_2(g)$$

25 cm^3 of hydrogen peroxide was titrated with a 0·1 mol l^{-1} acidified potassium permanganate solution and the results are shown in the table

Titration	Volume of potassium permanganate solution/cm³
1	16·2
2	15·7
3	15·9

(a) Why is the potassium permanganate solution acidified?   (1 Mark)

(b) Use the results to calculate the concentration of the hydrogen peroxide solution.   (3 Marks)

Answer sheet for Section A:

Higher Chemistry
Practice Papers for SQA Exams

Please select your answer using a single mark e.g.

A B C D

A B C D   A B C D

1. ☐ ☐ ☐ ☐   21. ☐ ☐ ☐ ☐
2. ☐ ☐ ☐ ☐   22. ☐ ☐ ☐ ☐
3. ☐ ☐ ☐ ☐   23. ☐ ☐ ☐ ☐
4. ☐ ☐ ☐ ☐   24. ☐ ☐ ☐ ☐
5. ☐ ☐ ☐ ☐   25. ☐ ☐ ☐ ☐
6. ☐ ☐ ☐ ☐   26. ☐ ☐ ☐ ☐
7. ☐ ☐ ☐ ☐   27. ☐ ☐ ☐ ☐
8. ☐ ☐ ☐ ☐   28. ☐ ☐ ☐ ☐
9. ☐ ☐ ☐ ☐   29. ☐ ☐ ☐ ☐
10. ☐ ☐ ☐ ☐   30. ☐ ☐ ☐ ☐
11. ☐ ☐ ☐ ☐   31. ☐ ☐ ☐ ☐
12. ☐ ☐ ☐ ☐   32. ☐ ☐ ☐ ☐
13. ☐ ☐ ☐ ☐   33. ☐ ☐ ☐ ☐
14. ☐ ☐ ☐ ☐   34. ☐ ☐ ☐ ☐
15. ☐ ☐ ☐ ☐   35. ☐ ☐ ☐ ☐
16. ☐ ☐ ☐ ☐   36. ☐ ☐ ☐ ☐
17. ☐ ☐ ☐ ☐   37. ☐ ☐ ☐ ☐
18. ☐ ☐ ☐ ☐   38. ☐ ☐ ☐ ☐
19. ☐ ☐ ☐ ☐   39. ☐ ☐ ☐ ☐
20. ☐ ☐ ☐ ☐   40. ☐ ☐ ☐ ☐

Exam 3

Chemistry   Higher

Practice Papers
For SQA Exams

Exam 3
Higher Level

Fill in these boxes:

Name of centre

Town

Forename(s)

Surname

You may use the Chemistry Higher and Advanced Higher Data Booklet.

SECTION A

Answers should be marked in HB pencil on a copy of the answer sheet on page 46 of this book.

SECTION B

1. Attempt all questions
2. Write you answers clearly, in ink
3. Write rough work in this book, then score it out clearly when you have written the fair copy.
4. The space available for answers does not indicate how much you should write. You do not have to fill all the available space.

Section A
Attempt all questions in this section.
Mark your answers on a copy of the answer sheet on page 46 of this book.

1. Which of the following substances has a covalent network structure but **does** conduct electricity?

   A   Sodium chloride
   B   Aluminium
   C   Silicon dioxide
   D   Graphite

2. Which pair of metals will produce an electron flow in the direction shown?

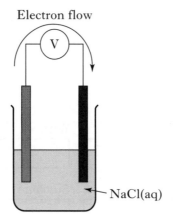

   A   Mg/Zn
   B   Zn/Mg
   C   Cu/Zn
   D   Ag/Zn

3. The corrosion of iron can be slowed by

   A   connecting the iron to the negative terminal of a battery
   B   sitting the iron in salty water
   C   connecting the iron to the positive terminal of a battery
   D   wrapping a piece of copper wire around the iron.

4. The same mass of zinc was dropped into two beakers both containing an excess of sulphuric acid, concentrations of 2 mol l^{-1} and 1 mol l^{-1} respectively. Which of the following would be the **same** for the two reactions?

   A   the time taken for the reaction to go to completion
   B   the initial rate of reaction
   C   the total mass lost
   D   the average rate of reaction.

5. Which of the following compounds has the highest molecular mass?

A $^{12}C^{16}O_2$
B $^{12}C^{17}O_2$
C $^{14}C^{16}O_2$
D $^{14}C^{17}O_2$

6. The potential energy diagram for a reaction is shown below.

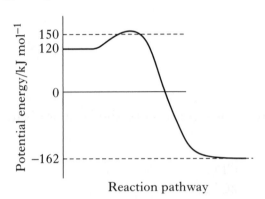

The enthalpy change for the forward reaction in kJ mol^{-1} is

A −42
B 155
C −282
D −182

7.

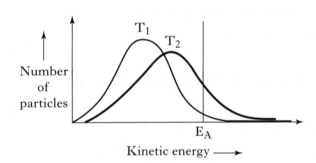

Which line in the table correctly interprets the above energy distribution diagram for a reaction as the temperature increases from T_1 to T_2?

	Activation energy (E$_A$)	Number of successful collisions
A	Unchanged	Increases
B	Unchanged	Decreases
C	Increases	Increases
D	increases	Decreases

8. What type of bond is broken when ice is melted?

    A   covalent
    B   polar covalent
    C   van der Waals'
    D   hydrogen

9. Which of the following elements is the least electronegative?

    A   Fluorine
    B   Chlorine
    C   Sodium
    D   Caesium

10. Which of the following reactions represents the first ionisation enthalpy of chlorine?

    A $Cl_2(l) \longrightarrow Cl^-(g)$
    B $Cl_2(g) \longrightarrow 2Cl^-(g)$
    C $Cl(g) \longrightarrow Cl^+(g)$
    D $Cl(g) \longrightarrow Cl^{2+}(g)$

11. An element melts at 64°C and reacts with water to produce a solution with a pH greater than 12. Which statement about the element is correct?

    A   The oxide of the element is acidic
    B   The element has a covalent network structure
    C   The oxide of the element is insoluble in water
    D   The element conducts electricity

12. Which of the following has the greatest number of molecules?

    A   0·1 g of hydrogen gas
    B   0·16 g of methane
    C   0·34 g of ammonia
    D   0·14 g of nitrogen

13. Which of the following has a covalent molecular structure?

    A   Argon
    B   Fullerene
    C   Calcium chloride
    D   Silicon dioxide

14. Three moles of oxygen is mixed with one mole of methane gas and ignited. What is the number of moles of gas in the resulting gas mixture at room temperature?

    A   1
    B   2
    C   3
    D   4

15. How many moles of ions are present in 0·2 moles of aluminium sulphate?

    A   0·2 moles
    B   0·4 moles
    C   0·6 moles
    D   1 mole

16. Listed below are some of the pollutant gases produced by cars. Which one of the gases is **not** the result of incomplete combustion?

    A   Carbon monoxide
    B   Carbon
    C   Nitrogen dioxide
    D   Hydrocarbons

17. In which process does the following reaction take place?

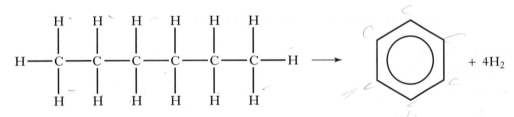

    A   dehydration
    B   hydration
    C   cracking
    D   reforming

18. Biogas refers to a gas produced by the biological breakdown of organic matter in the absence of oxygen. What is the main constituent of biogas?

    A   Hydrogen
    B   Propane
    C   Nitrogen
    D   Methane

19. Which of the following is an isomer of 3,3-dimethylbutan-1-ol?

    A $CH_3CH_2CH(CH_3)CH_2OH$
    B $CH_3CH(CH_3)CH(CH_3)CH_2OH$
    C $CH_3CH_2CH(CH_3)CH_2OH$
    D $CH_3CH(CH_3)C(CH_3)_2CH_2OH$

20. Alcohols can be used to produce alkenes. What type of reaction must be carried out to achieve this?

    A   dehydration
    B   hydrolysis
    C   condensation
    D   dehydrogenation

21. Paracetamol is a widely used over-the-counter pain reliever.

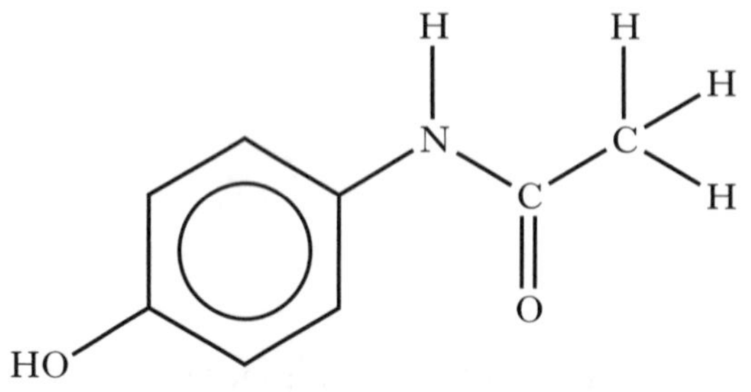

    What two functional groups are present in paracetamol?

    A   hydroxyl and carbonyl
    B   amide link and hydroxyl
    C   amine and carbonyl
    D   amine and carboxyl

22. Which of the following compounds does **not** represent a carbohydrate?

    A $C_{12}H_{22}O_{11}$
    B $(C_6H_{10}O_5)_n$
    C $C_3H_6O_2$
    D $C_6H_{12}O_6$

23. Which of the following would not react with acidified potassium dichromate solution?

    A   Butan-2-ol
    B   Butanone
    C   Methanol
    D   Methanal

24. What job does Ozone perform in the upper atmosphere?

    A   reflects ultraviolet radiation
    B   absorbs ultraviolet radiation
    C   absorbs CFCs
    D   removes greenhouse gases

25. The compound shown is an example of

A   a primary alcohol
B   a secondary alcohol
C   a tertiary alcohol
D   an alkanal.

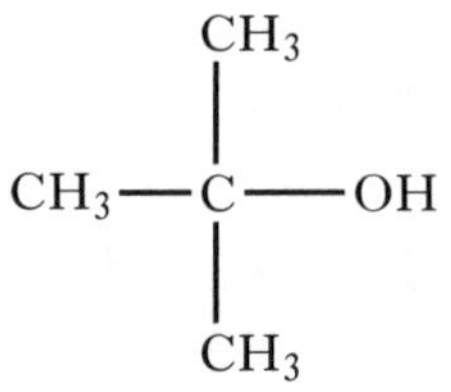

$$CH_3 - \underset{\underset{CH_3}{|}}{\overset{\overset{CH_3}{|}}{C}} - OH$$

26. During the formation of polyester what two functional groups combine during the polymerisation process?

A   hydroxyl and carbonyl
B   hydroxyl and carboxyl
C   amine and carboxyl
D   carbonyl and carboxyl

27. Which of the following statements about poly(ethene) and nylon is correct?

A   Both are formed by condensation polymerisation.
B   Both are formed by addition polymerisation.
C   Both have hydrogen bonding between their chains.
D   Both give off carbon dioxide when burned.

28. When an egg is heated, the protein it contains is denatured, causing it to change colour from colourless to white. During denaturing, what happens to the protein molecule?

A   It is hydrolysed
B   It changes shape
C   It is dehydrated
D   It is polymerised.

29. How many hydroxyl groups does glycerol contain?

A   0
B   1
C   2
D   3

30. Propene is an important feedstock for the production of plastics. It is produced from propane by following process.

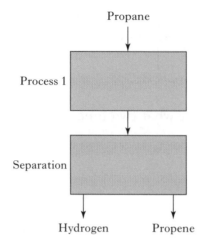

What is process 1?

A   distillation
B   evaporation
C   cracking
D   dehydration

31. Which of the following is produced by a batch process?

A   Iron in a blast furnace
B   Ammonia from the Haber process
C   Aspirin from salicylic acid
D   Nitric acid from the Ostwald process

32. The three equations shown below all involve displacement reactions between metals and metals oxides.

$Mg(s) + FeO(s) \longrightarrow MgO(s) + Fe(s) \Delta H = \mathbf{A} \text{ kJ mol}^{-1}$
$Fe(s) + CuO(s) \longrightarrow FeO(s) + Cu(s) \Delta H = \mathbf{B} \text{ kJ mol}^{-1}$
$Mg(s) + CuO(s) \longrightarrow MgO(s) + Cu(s) \Delta H = \mathbf{C} \text{ kJ mol}^{-1}$

What is the relationship between A, B and C according to Hess's Law?

A $A + B = -C$
B $A + B = C$
C $C + A = -B$
D $C + A = B$

33. When solid ammonium chloride is added to a dilute solution of ammonia which of the following ions will decrease in concentration to counteract this change?

A   Chloride
B   Ammonia
C   Ammonium
D   Hydroxide

34. One method used to produce methanol requires synthesis gas. The following equation shows the production of methanol from synthesis gas.

$$2H_2(g) + CO_2(g) \longrightarrow CH_3OH(g) \quad \Delta H = -91 \text{ kJ mol}^{-1}$$

Which line in the table shows the conditions that would cause the greatest increase in the amount of methanol produced?

	Pressure	Temperature
A	High	High
B	Low	Low
C	High	Low
D	Low	High

35. A loch has been polluted by acid rain and the pH is recorded at 4. Lime is added to the loch and over time the pH increases to 6. The concentration of H+ ions

A decreased by a factor of 2
B decreased by a factor of 100
C increased by a factor of 10
D increased by a factor of 100

36. The concentration of OH^- ions in a solution is $0 \cdot 01$ mol l^{-1}. What is the pH of the solution?

A 2
B 4
C 12
D 14

37. An unknown white solid is dissolved in water, causing the pH of the solution to increase. The unknown solid also reacts with hydrochloric acid to produce carbon dioxide gas. The solid could be which of the following?

A Sodium oxide
B Copper(II) carbonate
C Aluminium oxide
D Potassium carbonate

38. The ion electron equations for the oxidation of copper by nitric acid are shown

$$Cu(s) \longrightarrow Cu^{2+}(aq) + 2e^-$$
$$NO_3^-(aq) + 4H^+(aq) + 3e^- \longrightarrow NO(g) + H_2O(l)$$

How many moles of copper ions are oxidised by 1 mole of nitrate ions?

A $0 \cdot 33$
B $0 \cdot 67$
C $1 \cdot 50$
D $3 \cdot 00$

39. An atom of ^{227}Th decays by alpha emission to produce an atom of ^{211}Pb. How many alpha particles were released to produce this lead atom?

  A   1
  B   2
  C   3
  D   4

40. The following graph shows the radioactive decay of sodium-24. What is the half-life of this sample?

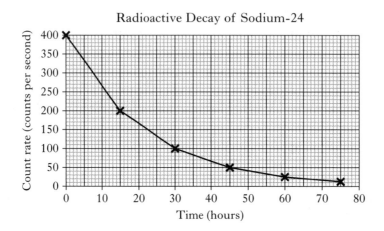

Radioactive Decay of Sodium-24

  A   10 hours
  B   15 hours
  C   60 hours
  D   200 hours

SECTION B
Write your answers clearly in ink.

1. (a) Complete the table below by adding the structure and type of bonding present in each of the following elements.

Element	Bonding and structure at room temperature
Sulphur	Discrete covalent molecular solid
Helium	
Carbon (diamond)	
Oxygen	
Potassium	

(2 Marks)

(b) Why do covalent molecular solids such as sulphur not conduct electricity?

(1 Mark)

2. The structure of a fat molecule is shown.

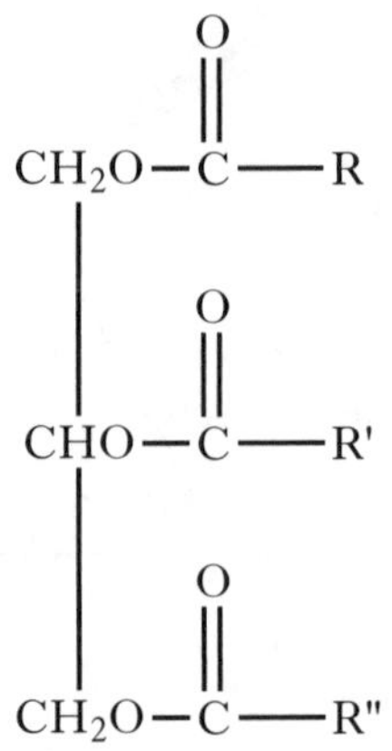

$$
\begin{array}{c}
O \\
\| \\
CH_2O-C-R \\
| \\
O \\
\| \\
CHO-C-R' \\
| \\
O \\
\| \\
CH_2O-C-R''
\end{array}
$$

(a) When the fat is hydrolysed fatty acids are obtained. The fatty acids are represented by R, R' and R" in the diagram. Name the other product of this reaction. (1 Mark)

(b) Fats are solid at room temperature, but oils are liquid. Why do fats have a higher melting point than oils (1 Mark)

(c) Due to their functional group, fats and oils can be classified as what type of compound? (1 Mark)

3. The potential energy diagram for a reaction is shown.

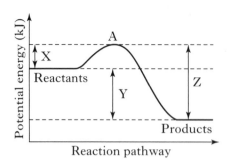

(a) What could be used to lower the values of X and Z but will have no effect on Y? (1 Mark)

(b) At point A an unstable arrangement of atoms is formed. What name is given to this arrangement? (1 Mark)

(c) Is the reaction exothermic or endothermic? (1 Mark)

4. An experiment to study the effect of concentration of potassium iodide on its rate of reaction with hydrogen peroxide was performed by students in the lab. They mixed 25 cm³ of KI with sodium thiosulphate solution, dilute sulphuric acid and starch solution. A few cm³ of hydrogen peroxide was added and the time taken for the colour change to occur was recorded.

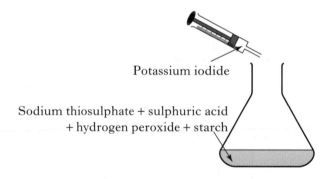

Potassium iodide

Sodium thiosulphate + sulphuric acid + hydrogen peroxide + starch

The experiment was repeated using different concentrations of potassium iodide. The results table for the experiment has not been completed.

(a) Complete the table of results by giving the values of X and Y. (2 Marks)

Volume of KI/cm³	Volume of water/cm³	Time/s	Rate /s⁻¹
25	0	**X**	0·0454
20	5	25	0·0400
15	10	33	**Y**
10	20	48	0·0208
5	30	100	0·0100

(b) What colour change would indicate that the reaction had finished? (1 Mark)

(c) Draw a graph of the volume of potassium iodide against the rate of reaction. (3 Marks)

(d) The method of dilution of the potassium iodide ensures that the total volume is kept constant. Why is this important in the experiment? (1 Mark)

5. The enthalpy of combustion of ethanol can be measured using a bomb calorimeter like the one shown.

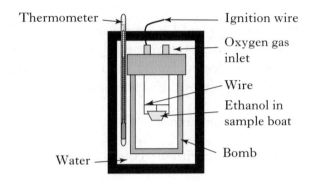

(a) The results obtained for the enthalpy of combustion of ethanol using the bomb calorimeter are higher than those obtained in the school lab. One reason for this is that heat can be lost to the surroundings in the lab. Give another possible reason for the different results. (1 Mark)

(b) When the experiment was performed in the calorimeter, it was found that 0·92 g of ethanol resulted in the temperature of 400 cm³ of water increasing from 18·2°C to 34·3°C. Calculate the enthalpy of combustion of ethanol. (3 Marks)

6. Green salt (uranium tetrafluoride) is used to produce fuel for nuclear power stations. It is produced from uranium ore.

(a) Uranium can be extracted from green salt in a redox reaction with magnesium metal.

$$2Mg + UF_4 \longrightarrow 2MgF_2 + U$$

Give another name for this type of reaction. (1 Mark)

(b) For this reaction to take place, the UF_4 must be in the molten state. Therefore the reaction is carried out a temperature of over 1100°C. The reaction is carried out in an argon atmosphere. Give a reason why the reaction is not carried out in air. (1 Mark)

(c) Uranium hexafluoride is a compound produced from uranium tetrafluoride. Both compounds are radioactive because they contain the same isotope. How would the half-life of UF_6 compare to that of UF_4? (1 Mark)

(d) Uranium hexafluoride can also be used as a fuel for nuclear power plants. Listed below are some of the properties of UF_6. Suggest the type of bonding present in Uranium hexafluoride. (1 Mark)

Properties of UF6.

Appearance	Colourless solid
Density	$5 \cdot 09$ g/cm^3
Melting point	$64 \cdot 8°C$
Hazard	Very toxic

7. The reaction between hydrochloric acid and magnesium can be used to calculate the molar volume of hydrogen. A student performed this experiment by reacting 0·20 g of magnesium with excess hydrochloric acid. The hydrogen gas was collected using the downward displacement of water into a measuring cylinder

(a) Write a balanced equation for the reaction. (1 Mark)

(b) Draw a diagram of how this experiment could be performed in the lab (2 Marks)

(c) When the reaction was complete the student had collected 200 cm^3 of hydrogen gas in the measuring cylinder. Calculate the molar volume of hydrogen gas at this temperature and pressure. (2 Marks)

(d) The student ensured that the acid used in this experiment was in excess. Why was this essential? (1 Mark)

8. Ionic compounds which fit the formula XY, where X is the metal ion and Y is the non-metal ion take up one of two possible cubic arrangements. The arrangement of ions is determined mainly by the radius ratio of the ions involved. These are given in the Data Booklet.

$$\text{Radius Ratio} = \frac{\text{Radius of positive ion}}{\text{Radius of negative ion}}$$

The arrangements for sodium chloride and caesium chloride are shown below.

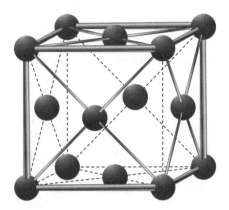

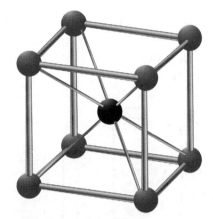

Sodium chloride
Radius ratio = 0·52
Face-centred cubic arrangement

Caesium chloride
Radius ratio = 0·96
Body-centred cubic arrangement

(a) If the radius ratio is above 0·8 then a compound will have a structure similar to sodium chloride. If it is below 0·8 then it will have a structure similar to caesium chloride. Calculate the radius ratio of magnesium oxide and state which of the two arrangements – face-centred cubic or body-centred cubic – it will have. (1 Mark)

(b) Potassium oxide has neither body or face centred cubic structures. Suggest a reason for this. (1 Mark)

9. A student performed electrolysis on two different solutions. In the first experiment he electrolysed a solution of silver(I) nitrate and produced 0·54 g of silver.

(a) At which electrode would the silver be formed? (1 Mark)

(b) If he passed the same amount of electricity through a sulphuric acid solution, what volume of hydrogen would he produce? Take the molar volume to be 23·8 litres at this temperature and pressure (3 Marks)

10. The properties of hydrochloric acid and methanoic acid are compared in the table below

Property	Methanoic Acid	Hydrochloric Acid
Conductivity	Low	High
pH	4	1

Using methanoic acid and hydrochloric acid as examples, explain the differences in both pH and conductivity between 1 mol l^{-1} solutions of a strong and a weak acid. You should include equations in your explanation. (3 Marks)

11. An enzyme that is present in potatoes can be used to catalyse the breakdown of hydrogen peroxide into water and oxygen as shown by the following equation

$$2H_2O_2(aq) \longrightarrow 2H_2O(l) + O_2(g)$$

The rate of decomposition of hydrogen peroxide can be observed using the following apparatus.

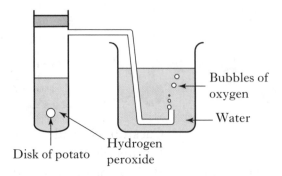

(a) Describe fully how the apparatus shown can be used to investigate the effect of temperature on the rate of decomposition of hydrogen peroxide.
(2 marks)

(b) The graph below illustrates the effect of temperature on enzyme catalysed reactions.

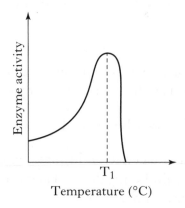

Suggest a value for temperature T_1 (1 Mark)

(c) A method used to explain how enzymes work is the 'Lock and Key' method. Complete the diagram below to illustrate this method. (1 Mark)

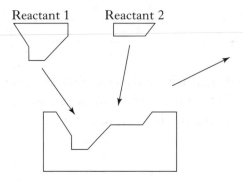

12. Acrylonitrile is an important monomer for the production of plastics. It has the formula CH_2CHCN. It is used principally in the production of carbon fibre.

    (*a*) Draw the full structural formula of acrylonitrile.   (1 Mark)

    (*b*) What type of polymerisation takes place to convert the monomer acrylonitrile into a polymer?   (1 Mark)

    (*c*) Acrylonitrile can be reduced to form CH_2CH_2CN. Complete and balance the ion-electron equation for the reduction reaction.   (1 Mark)

$$CH_2CHCN \longrightarrow (CH_2CH_2CN)_2 + OH^-$$

13. August Kekulé was an organic chemist born in Germany in 1829. His work involved predicting the structure of organic compounds. One of the structures he worked on was the structure of benzene. The structure that he proposed for benzene is shown below.

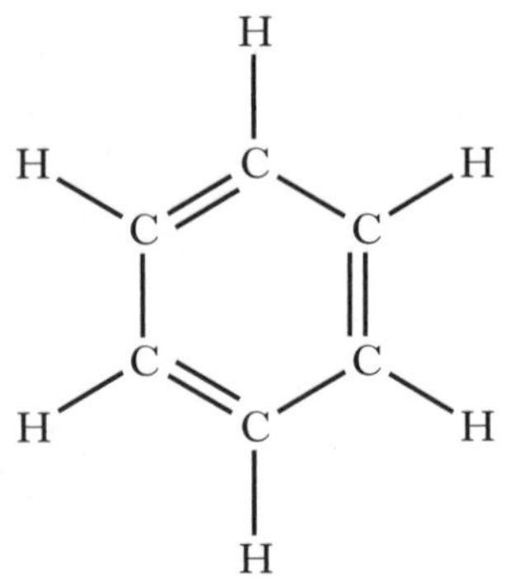

    (*a*) Describe a chemical test that would suggest that the structure proposed by Kekulé is incorrect.   (1 Mark)

    (*b*) Draw the correct structure of benzene.   (1 Mark)

    (*c*) In recent years benzene has been added to unleaded petrol. Suggest a reason for this.   (1 Mark)

    (*d*) The equation for the formation of benzene from carbon and hydrogen is shown below.

$$6C(s) + 3H_2(g) \longrightarrow C_6H_6(l)$$

    Use the enthalpies of combustion from your data book to calculate the enthalpy of formation of benzene   (3 Marks)

14. Shown below are three organic molecules

$$CH_3-HC{=}CH-CH_3 \longrightarrow \qquad CH_3CH_2-CHClCH_3 \longrightarrow \qquad CH_3CH_2-CH(OH)CH_3$$

Molecule A **Molecule B** **Molecule C**

(*a*) Give the systematic name of each of the three compounds. (3 Marks)

(*b*) Name the compound that would be required to convert molecule A into molecule B. (1 Mark)

(*c*) Molecule C can be oxidised further to produce a new compound. Draw the full structural formula and give the systematic name of the product. (2 Marks)

(*d*) Another molecule was tested with a reagent, with which it formed a silver mirror. Name the reagent used. (1 Mark)

(*e*) Name another reagent that could be used to test the reducing properties of the molecule. (1 Mark)

Worked Answers

SECTION A

Question	Answer	Hint
1	B	Ionic substances only conduct when molten or in solution.
2	A	Ferroxyl indicator turns blue with Fe^{2+} ions. The salt water and the direction of electron flow indicate that the iron is corroding producing Fe^{2+} ions
3	C	Sodium (2,8,1) loses an electron to form sodium ions (2,8). Fluorine (2,7) gains an electron to form fluoride ions (2,8). Neon has the electron arrangement of 2,8.
4	B	There are several ways to do this calculation. The most common is the $P \times V \times C_{(acid)} = P \times V \times C_{(alkali)}$ method. **Power** = number of H^+ or OH^- ions. **Volume** in litres. **Concentration** in mol l^{-1}.
5	C	Think about it! If there are two isotopes of Hydrogen then there are three possible types of molecule: ^1H-^1H, ^1H-^2H, and ^2H-^2H. This is a common question but a difficult one.
6	C	Note the word average. The equation used to calculate **average** rate is: Average Rate = Change in concentration ÷ Change in time ($0.15 \div 20$)
7	C	This is easier than it first appears. If 10 g produces 310 kJ then 60 g produces six times the energy, 1860 kJ. No need to do $\Delta H = cm\Delta T$
8	C	Take your time with these and narrow down the options. Activation energy is energy difference from the reactants to the peak of the graph (80 to 180) which is 100. The enthalpy change shows an overall loss of 40 kJ. Energy loss means the reaction is exothermic.
9	D	Look for the compound in which the elements have the biggest difference in electronegativity, or are the furthest apart in the periodic table.
10	C	First ionisation is the energy required to remove **one mole of electrons** from one mole of atoms in the **gaseous state**.
11	B	Learn all the substances that exist as covalent networks. It is a regular question in exams.
12	B	It helps to draw the full structural formula out on spare paper. The presence of an –OH group is a sure sign that hydrogen bonding will occur
13	A	**'Like Dissolves Like'** Covalent substances dissolve in covalent solvents such as tetrachloromethane (carbon tetrachloride).
14	D	Write out the balanced equation for the combustion of methane and you will see that 1 mole of methane produces 2 moles of water. So 2 moles of methane would produce 4 moles of water.
15	D	This is easier than it first appears. The greatest number of moles gives the greatest number of molecules. All you have to do is work out moles using: Moles = mass ÷ formula mass.
16	B	1 mole of sodium oxide, 62 g, contains 3 moles of ions ($3 \times 6.02 \times 10^{23}$) because of the formula $(Na^+)_2 O^{2-}$. So 31 g is half a mole, so that must contain $1.5 \times 6.02 \times 10^{23}$ ions.
17	B	Learn all the types of reaction because they appear more than once in most exams. A is dehydration, C is cracking and D is addition/hydrogenation.
18	A	Don't panic! The molecule may appear very complicated but all you have to do is identify the two functional groups. Remember that COOH is a carboxyl and ester is a COO!

SECTION A (continued)

Question	Answer	Hint
19	A	To do this question correctly, it is essential to draw out the full structural formulas of the compounds and any possible isomers on spare paper.
20	D	Again, you have to draw out the full structural formulas to answer this question correctly. A tertiary alcohol's OH group is attached to a carbon atom which has **3 carbon** atoms attached to it.
21	B	Only primary alcohols can be oxidised to produce carboxylic acids. The compound is butanoic acid so the primary alcohol used to produce it is butan-1-ol.
22	D	The formula of benzene is C_6H_6 and ethyne is C_2H_2.
23	A	Synthesis gas appears in **ALL** exam papers. You must know what it is, how it is made and what it is used for.
24	C	All exam papers include a question on modern plastics and their properties.
25	A	Polymerisation is stopped by adding a monomer that has only **one** functional group.
26	A	Fats are produced from fatty acids and glycerol. Glycerol has the systematic name propan-1,2,3-triol.
27	B	Hydrogenation reduces the number of carbon to carbon double bonds and therefore increases the melting point.
28	B	Enzymes are classed as proteins.
29	C	Methane is produced naturally and is therefore a raw material. The others are produced by chemical processes – Iron (blast furnace), Ammonia (Haber process), Benzene (reforming)
30	A	The value of ΔH_1 is equal to X and ΔH_2 combined. These questions are fairly simple but pay close attention to the direction of the arrow because it determines whether the reaction is exothermic or endothermic so you may have to reverse it on some occasions.
31	C	Catalysts **do not** effect the position of equilibrium – they simply allow equilibrium to be established quicker.
32	C	Exothermic reactions favour a low temperature. The equation shows that there is only 1 mole of gaseous products and 2 moles of gaseous reactants. A high pressure favours the side with the lowest gas volume so a high pressure will favour the production of ICl_3 on this occasion.
33	C	Sulphuric, hydrochloric and nitric are all strong acids.
34	C	Manipulate the ionic product of water equation to calculate the pH. $[H^+] \times [OH^-]$. $[H^+]=10^{-14} \div [OH^-] = 10^{-14}$ therefore $[H^+]=10^{-14} \div [OH^-] = 10^{-12}$ so pH is 12
35	B	Solutions of strong and weak acids, with the same concentration and volume, require the same amount of alkali to neutralise them.
36	A	This is a tricky question. You must remember to balance the equation first by putting a 2 in front of the ClO_3^- before adding $6H_2O$ to the other side and then $12H^+$ ions to the left hand side.
37	B	Potassium is being oxidised to potassium ions by hydrogen to form potassium hydride
38	B	Isotopes have the same atomic number but different mass numbers
39	A	The half-life of a radioactive sample cannot be changed but the intensity of the radiation would be different.
40	A	Fusion is the combining of small particles to form a heavier particle.

SECTION B

Question	Answer	Hint
1(a)	2,2,4-trimethylpentane	Find the longest chain of carbons and number the branches to give the lowest numbers. For example 2,4,4-trimethylpentane is incorrect.
1(b)	Cyclic molecules or aromatic molecules	The octane number is a measure of how efficiently petrol burns.
2(a)	Low density/lightweight	Learn the properties of the following recently-developed plastics as they appear regularly in exams: Poly(ethenol), Kevlar, biopol, Poly(vinylcarbazole), Poly(ethyne), low density poly(ethene)
2(b)	Amine group	It is essential to learn all the homologous series and their functional groups.
2(c)	Condensation Polymerisation	Condensation polymerisation is the reaction in which many monomers combine to form a polymer and another **small molecule**. That small molecule however doesn't have to be water. In this case it is hydrogen chloride.
3(a)	$^{100}_{43}Tc$	All you have to do is add a neutron to the mass of 99.
3(b)	$^{100}_{43}Tc \longrightarrow \ ^{100}_{44}Ru + \ ^{0}_{-1}e$	Due to the negative atomic number of beta emission the atomic number increases.
3(c)	12·5g	The mass would half every 16s so if it is left for 48s it would halve three times: $100 \longrightarrow 50 \longrightarrow 25 \longrightarrow 12.5$
4(a)		Always check that each element in the compound has the correct number of bonds i.e. Carbon = 4 bonds, Nitrogen = 3 bonds and Hydrogen = 1 bond. You might not get it right first time so use spare paper.
4(b)	X = 4, Y = 9, Z = 12	A difficult equation to balance.

SECTION B (continued)

Question	Answer	Hint
4(c)	+53 kJ mol^{-1}	Top Tip for Hess's Law questions 1. Remember to balance the equations. 2. Remember to reverse the enthalpy sign when reversing an equation. 3. Remember to multiply the enthalpy value when multiplying an equation. Working You will need the balanced equations for the combustion of carbon and hydrogen. **Equation 1** $C(g) + O_2(g) \longrightarrow CO_2(g)$ $\Delta H = -394$ kJ mol^{-1} **Equation 2** $H_2(g) + \frac{1}{2}O_2(g) \longrightarrow H_2O(g)$ $\Delta H = -286$ kJ mol^{-1} **Equation 3** $4CH_3NHNH_2(l) + 2\frac{1}{2}O_2(g) \longrightarrow CO_2(g) + N_2(g) + 3H_2O(l)$ $\Delta H = -1305$ kJ mol^{-1} To match the target equation, equation 2 has to be multiplied by three and equation 3 must be reversed. The enthalpies can then be combined.
5(a)	Clear labelled diagram showing a water bath and paper towel condenser around the test tube and Sulphuric acid catalyst must be mentioned and Mention of sodium hydrogen carbonate solution to neutralise the acid	It is essential to learn all the PPAs in detail. There will be at least two PPA-based questions in each exam. The PPA covered here is from Unit 2: 'Preparation of an Ester'
5(b)	No naked flames	A water bath must be used to heat the flammable liquids used in this experiment.

SECTION B (continued)

Question	Answer	Hint
5(c)		Always check the number of bonds that each carbon atom forms is four.
5(d)	64% Partial marks for Moles of ester produced Theoretical mass of ester Percentage yield	Top Tip for percentage yield questions 1. Remember that all esterification reactions are a 1 to 1 ratio of alcohol to ester. So there is no need to balance the equation. Worked answer. Step 1 Work out the moles of alcohol using Moles = mass ÷ formula mass. $(9 \div 46 = 0 \cdot 196$ moles$)$ Step 2 The ratio is 1:1 so $0 \cdot 196$ moles of ethanol would produce $0 \cdot 196$ moles of ethyl ethanoate. (1 mark) Step 3 Calculate the theoretical mass of ethyl ethanoate produced using Mass = moles × formula mass $(0 \cdot 196 \times 88 = 17 \cdot 2g)$ (1 mark) Step 4 Calculate the percentage yield using % yield = Actual mass ÷ Theoretical mass × 100 $(11 \div 17 \cdot 2 \times 100 = 64\%)$ (1 mark)

SECTION B (continued)

Question	Answer	Hint
6(a)	Lead is in excess Partial marks for: Calculating the moles of both reactants establishing which reactant is in excess	Top Tip for questions involving $N = C \times V$ Remember that the volume must be in litres. Change cm^3 into litres by dividing by 1000. Step 1 Establish the molar ratio. 1 mole of lead reacts with two moles of hydrochloric acid. Step 2 Calculate the moles of each reactant using Moles = mass ÷ formula mass and $N = C \times V$ Moles of lead = $(10\cdot36 \div 207\cdot2 = 0\cdot05$ moles) Moles of acid = $(1 \times 0\cdot05 = 0\cdot05$ moles) (1 mark) Step 3 Use the ratio to establish the moles of each reactant required e.g. 1 mole of Pb reacts with two moles of HCl so $0\cdot05$ moles of Pb would react with $0\cdot1$ moles of HCl. This means that the Pb is in excess. This is the most difficult step and requires some thought. (1 mark)
6(b)	$0\cdot05$ g	Using the balanced equation, $0\cdot05$ moles of acid (don't use lead because it is in excess!) reacts to produce $0\cdot1$ moles of hydrogen gas. Then use Mass = moles × formula mass to get the answer.
6(c)	— Water	Markers assess these questions using 'Will the experiment work as drawn?'
7(a)	The carbonyl group is on the end carbon of aldehydes.	Learn all the homologous series and their functional groups.
7(b)	Blue to orange/brick red	Always remember to give the complete colour change. Don't just say that it 'went orange'. Questions 7(b) and 7(c) are based on the PPA 'Oxidation of Aldehydes and Ketones'.
7(c)	Acidified potassium dichromate or Tollen's reagent	Oxidising agents and their colour changes will be in all exams.
7(d)	Butanoic acid	Remember to include the carbon atom in the functional group when naming carboxylic acids.

SECTION B (continued)

Question	Answer	Hint
8(a)		Experiment 1 would produce the same volume of gas as experiment 2 but it would take longer to do this. Experiment 3 would produce more gas because of the higher concentration of the H_2O_2.
8(b)		Markers assess these questions using 'Will the experiment work as drawn?'
9(a)	They only have 1 electron	This is an Advanced Higher Chemistry question that is used in Higher as problem solving. They are difficult questions and will test your problem solving skills.
9(b)		Use the diagram in question 9(a) as your guide
10(a)	Sulphur ions (2,8,8) have one more energy level than magnesium ions (2,8)	Periodic trends are fairly easy to learn but you must also be able to explain the trends.
10(b)	Calcium has a greater nuclear charge.	Periodic trends are fairly easy to learn but you must also be able to explain the trends.

SECTION B (continued)

Question	Answer	Hint
11	Ethanol molecules are held together by hydrogen bonds due to the polarised –O–H group (1 mark). This occurs due to the difference in electronegativity between the hydrogen and oxygen atoms. (1 mark). Propane molecules are held together by van der Waals' forces which are much weaker than hydrogen bonds (1 mark). Van der Waals' are due to momentary displacement of electrons between atoms (1 mark).	In this type of question, you must give as much information as possible in your answer. You are trying to tell the examiner <u>everything</u> you know about bonding!
12(a)	Strong alkali and a weak acid	pH of salts are relatively straightforward: Strong acid + Strong alkali = Neutral salt Strong acid + Weak alkali = Acidic salt Weak acid + Strong alkali = Alkaline salt Weak acid + Weak alkali = Neutral salt
12(b)	Magnesium carbonate is produced which is insoluble in water. This can be easily removed by filtration	These types of question are difficult and require some careful thinking before you will obtain the answer.
12(c)	$Ca(OH)_2 + 2NH_4Cl \longrightarrow 2NH_3 + CaCl_2 + 2H_2O$	A tricky one to balance. Always leave the hydrogen and oxygen atoms until last.
13(a)	$2CH_3OH + 3O_2 + \longrightarrow 2CO_2 + 4H_2O$	The CH_3OH equation must be multiplied by two to cancel out the $12e^-$ in the other equation.
13(b)	Methanol can cause engine corrosion Methanol also produces CO_2 on combustion	Alternative fuels are very topical at present and because of this they appear regularly in exams. Learn the advantages of and disadvantages of each.
13(c)	0·11 litres Partial marks for:- Using Q= I x T to calculate the quantity of electricity used. Establishing that 1 mole of Hydrogen requires 19 300 C of electricity. Calculating the volume of hydrogen produced	Top Tips for questions involving $Q = I \times T$ 1. Remember that the time must be in seconds 2. Remember to do the ion electron equation to establish how many electrons are required for 1 mole. 3. Remember that 96 500 C are required for each electron in the equation. Worked Answer <u>Step 1</u> Calculate the quantity of electricity passed using $Q = I \times T$ ($0·5$ A $\times 1800$ s $= 900$ C) <u>Step 2</u> Use the equation to calculate the quantity of electricity required for 1 mole of Hydrogen ($2 \times 96\ 500$ C $= 193\ 000$ C) <u>Step 3</u> Use this to calculate the number of moles of hydrogen produced using Moles = actual ÷ theoretical ($900 \div 193\ 000 = 4·66 \times 10^{-3}$ moles) <u>Step 4</u> Multiply the number of moles by the molar volume ($4·66 \times 10^{-3} \times 24 = 0·11$ litres)

SECTION B (continued)

Question	Answer	Hint
14(a)	$4KMnO_4(s) \longrightarrow 2K_2O(s) + 4MnO_2(s) + 3O_2(g)$	
14(b)(i)	Repeat the experiment to obtain an average. Add drop by drop near end-point	The first or rough titration should never be included when calculating an average.
14 (b)(ii)	3 mol l^{-1}	Step 1 Establish the molar ratio. 5 moles of Fe^{2+} ions react with 1 mole of MnO_4^- Ions. Step 2 Calculate the moles of permanganate ions using $N = C \times V$ ($0.2 \times 0.025 = 5 \times 10^{-3}$) Step 3 Use the ratio to establish the moles of Fe^{2+} ions used ($5 \times 10^{-3} \times 5 = 0.025$ moles of Fe^{2+}) Step 4 Calculate the concentration of the iron(II) sulphate solution using $C = N \div V$. ($0.025 \div 0.00835 = 3$ mol l^{-1}.
15(a)	Transition metal	Always read the questions carefully especially the words in bold. If you didn't read it carefully you may have answered platinum.
15(b)	A heterogeneous catalyst is in a different state to the reactants	You should also be able to explain how heterogeneous catalysts work.
15(c)	It will increase	Low temperatures favour exothermic reactions.
16(a)	-300 kJ mol^{-1} Partial marks for:- Using $\Delta H = cm\Delta T$ to calculate the enthaply change for the reaction. Calculating the moles of ethanol used. Calculating the enthalpy of combustion of 1 mole of ethanol.	Top Tips for questions involving $\Delta H = cm\Delta T$ 1. Remember that 100 cm^3 of water is 0·1 kg 2. Remember that all combustion reactions are exothermic and ΔH should therefore have a negative charge Worked Answer Step 1 Calculate the enthaply change using $\Delta H = cm\Delta T$ ($4.18 \times 0.1 \times 10 = -4.18$ kJ mol^{-1}) Step 2 Calculate the moles of ethanol used using Moles = mass ÷ formula ($0.64 \div 46 = 0.014$ moles) Step 3 Use this to calculate the enthalpy for 1 mole using 0·014 moles produces -4.18 kJ mol^{-1} then 1 mole produces ($1 \div 0.014 \times -4.18 = -300$ kJ)
16(b)	Heat lost to surroundings/incomplete combustion/ evaporation of ethanol	Another PPA question. You must learn them in detail!
17(a)	Stage 5	This is another difficult question based in Advanced Higher. You must study the diagram carefully. The arrow on the X-axis shows the potential energy increasing so a release of energy would go down the scale.
17(b)	$Na(s) + \frac{1}{2}Cl_2(g) \rightarrow Na^+Cl^-(s)$	

SECTION A

Question	Answer	Hint
1	C	A straightforward one to get you started.
2	C	This is a complicated way of asking if a compound is soluble or not in water.
3	C	Magnesium (2,8,2) loses two electrons to form magnesium ions (2,8). Fluorine (2,7) gains an electron to form fluoride ions (2,8). Notice that questions 1, 2 and 3 all have the same answer. There is never a pattern to the answers so never change an answer because you think that there is one!
4	D	This is a common question but a difficult one. Firstly work out the formulas of the two compounds. Na^+Br^- has 0·6 moles of Br^- and therefore must have 0·6 moles of Na^+. $(Na^+)_2CO_3^{2-}$ has 0·4 moles of CO_3^{2-} and therefore has 0·8 moles of Na^+ (due to the formula) 0·6 + 0·8 = 1·4 moles of Na^+
5	D	Only an alkali will increase the pH of water and you should remember that ammonia is an alkali.
6	C	A tricky question that requires some thought. Remember to read the question carefully and work out E_A for the reverse reaction.
7	A	Learn the enthalpy statements for Combustion, Solution and Neutralisation.
8	D	A small increase in temperature increases the kinetic energy of particles which increases the number of successful collisions, i.e. the number of collisions with energy greater than or equal to the activation energy. Temperature does not affect activation energy.
9	A	In Practice Paper 1, question 11 asked about covalent networks. If you have learned the network substances then all other covalent substances are molecular.
10	B	Compounds in which the elements have the biggest difference in electronegativity have the most ionic character. The electronegativity of selected elements can be found on page 10 of your Data booklet.
11	B	Learn the properties of covalent networks, ionic, covalent molecular and metallic substances
12	B	Methane and carbon tetrachloride are non-polar because they are symmetrical. Carbon dioxide is non-polar because it is linear. The bonding section in Unit 1 is difficult and requires a lot of work to understand it fully.
13	A	80 g (2 moles) of argon contains $2 \times 6 \cdot 02 \times 10^{23}$ atoms. Oxygen is diatomic so 1 mole (32 g) contains $2 \times 6 \cdot 02 \times 10^{23}$ atoms of oxygen.
14	D	Sodium sulphide has the formula $(Na^+)_2S^{2-}$. So 2 moles of $(Na^+)_2S^2$ has six moles of ions $(2 \times Na^+$ and $1 \times S^{2-} \times 2$ moles)
15	A	131·3 g of xenon contains $1 \times 6 \cdot 02 \times 10^{23}$ atoms of xenon.
16	B	1 mole of magnesium chloride contains 3 moles of ions $(3 \times 6 \cdot 02 \times 10^{23})$ because of the formula $Mg^{2+}(Cl^-)_2$. So 47·65 g is half a mole, so that must contain $1 \cdot 5 \times 6 \cdot 02 \times 10^{23}$ ions.
17	C	This question is relatively straightforward when you think about it. Due to the molar ratio, 1 mole of methane reacts with 2 moles of oxygen so 20 cm³ of methane would react with 40 cm³ of oxygen this leaves 60 cm³ of oxygen. The reaction also produces 20 cm³ of CO_2 so the total gas volume is 80 cm³ (60 cm³ + 20 cm³). Remember not to include water as it is a liquid in this question.

SECTION A (continued)

Question	Answer	Hint
18	C	Straight chain hydrocarbons do not improve the octane number of petrol.
19	D	A ketone's carbonyl group is on a carbon that is **not** at the end of the chain.
20	B	You have to draw out the full structural formulas to do this question correctly.
21	B	**Reduction is the loss of oxygen or the gain of hydrogen.** Count the number of hydrogen and oxygen atoms on each side of the arrow to see which one fits the statement. The other three options are all oxidation
22	C	Identify the weak acid. Strong alkali and weak acid produces an alkaline salt.
23	B	If you draw out the full structural formulas these types of questions can appear slightly easier.
24	B	Chlorofluorocarbons are better known as CFCs.
25	C	Resins have 3-D structure with cross linking between the chains.
26	B	Glycerol has the systematic name propane-1,2,3-triol. You should also be able to draw out the full structural formula of glycerol.
27	B	Draw out the full structural formula of the peptide link and it will become clear which bond is broken during hydrolysis,
28	D	Fats are saturated.
29	C	The chemical industry section of unit three can often be overlooked. This section must be studied in detail.
30	B	The third equation must be reversed.
31	B	A reaction is at equilibrium when the concentration of the reactants and products remains constant but not necessarily equal.
32	C	Pressure won't affect a reaction when there are the same number of moles of gaseous reactants as there are gaseous products.
33	C	A pH of four indicates a hydrogen ion concentration of 1×10^{-4} mol l^{-1}.
34	A	A pH change of 1 is a tenfold change in the hydrogen ion concentration.
35	A	Solutions of strong acids have a lower pH and a higher conductivity than a solution of a weak acid of the same concentration.
36	A	The acid part of the molecule will lose an H$^+$ ion most easily in solution.
37	B	Sn^{2+} is becoming Sn^{4+} which shows a loss of two electrons.
38	A	All displacement reactions are examples of redox reactions.
39	C	Divide the number of neutrons by the atomic number to calculate the ratio. $(126 \div 84)$
40	A	Alpha radiation is stopped by paper, beta is stopped by aluminium foil and gamma is stopped by lead.

SECTION B

Question	Answer	Hint
1(a)	Electronegativity is a measure of the attraction that an atom has for bonded electrons	Periodic trends are easy to learn but you also have to explain the trends. Electronegativity appears in all exams but usually in a question alongside polar covalent bonding.
1(b)	Atomic size decreases	You should also be able to explain why this is the case.
1(c)	The first ionisation of sodium is lower because it has an extra energy level (1 Mark). This means that the electron that is removed from lithium is closer to the positively charged nucleus and therefore more energy is required to remove it (1 Mark)	'Explain fully' is an important part of this question. It means that you must give as much detail as you can to obtain the full marks. One word answers will not work.
2(a)	Octane	A simple question from Standard Grade.
2(b)	Any branched alkane with the molecular formula C_8H_{18}.	Always double check that you have the correct number of carbon and hydrogen atoms in your formula.
2(c)	Conducts electricity when exposed to light. (photoconductivity)	Learn the properties of the following recently developed plastics as they appear regularly in exams:- Poly(ethanol), Kevlar, biopol, Poly(vinylcarbazole), Poly(ethyne), low density poly(ethane)
3(a)	Nitrogen and carbon dioxide	
3(b)	Heterogeneous catalysts work by the method of adsorption (1 Mark). The reactant molecules are adsorbed onto the active sites of the catalyst and held at favourable angles to allow collisions with other reactants. The product molecules than leave the active sites. (1 Mark)	Please note the detail that is required to gain the 2 marks. The key words that should be included in your answer are highlighted.
4(a)	The volume of Hydrogen gas collected, the current used and the time during which the solution was electrolysed.	It is essential to learn all the PPAs in detail. There will be at least two PPA based questions in each exam. The PPA covered here is from Unit 3 – 'Quantitive Electrolysis'

SECTION B (continued)

Question	Answer	Hint
4(b)	2·25g	Top Tips for questions involving $Q = I \times T$ 1. Remember that the time must be in seconds 2. Remember to do the ion-electron equation to establish how many electrons are required for 1 mole. 3. Remember that 96 500 C are required for each electron in the equation. Worked Answer Step 1 Calculate the moles of hydrogen gas collected using Moles = mass ÷ formula mass (0·25 ÷ 2 = 0·125 moles). Remember that hydrogen is diatomic! Step 2 Use the equation to calculate the quantity of electricity required for 1 mole of hydrogen (2 × 96 500 C = 193 000 C) Step 3 Calculate the quantity of electricity required to produce 0·125 moles of hydrogen. (193 000 × 0·125 = 24 125 C) **(1 mark)** Step 4 Use the equation to calculate the quantity of electricity required for 1 mole of aluminium (3 × 96 500 = 289 500 C) Step 5 Calculate the number of moles of aluminium produced from 24 125 C of electricity. (24 125 ÷ 289 500 = 0·083 moles) **(1 mark)** Step 6 Calculate the mass of aluminium produced using mass = moles × formula mass (0·083 × 27 = 2·25 g) **(1 mark)**
5(a)	Exothermic reaction	You deserved an easy question after that last one!
5(b)	−35·9 kJ	Top Tips for questions involving $\Delta H = cm\Delta T$ 1. Remember that 100 cm³ of water is 0·1 kg 2. Remember that all combustion reactions are exothermic and ΔH should therefore have a negative charge Worked Answer Calculate the enthalpy change using $\Delta H = cm\Delta T$ (4·18 × 0·227 × 37·8 = −35·9 kJ)

SECTION B (continued)

Question	Answer	Hint
6(a)	Purple to colourless	It is essential to learn all the PPAs in detail. There will be at least two PPA based questions in each exam. The PPA covered here is from Unit 1 – 'Effect of Temperature on Reaction Rate'
6(b)	The reaction is too slow at temperatures below 40°C and too fast at temperatures above 80°C.	
6(c)	E_A	A 10°C increase in temperature can double the rate of a reaction by increasing the kinetic energy of the particles involved. Increasing the temperature has no effect on the activation energy.
7(a)	Stage 1 is precipitation (1 mark) and stage 2 is neutralisation (1 mark)	Questions involving flow diagrams are always difficult questions and require a lot of thinking to answer correctly. Try writing out the equations for stage 1 and stage 2. It may make the answer easier to find.
7(b)	Electrolysis	Electrolysis can be used to separate an ionic solution using electricity.
7(c)	An arrow going from the chlorine produced in stage three to the chlorine entering stage 2	Flow diagram questions often involve a question based on how to make the process shown more economical.
8(a)	Benedict's reagent	You should be able to remember this from Standard Grade.
8(b)	Weak acids are only partially ionised in aqueous solution.	

SECTION B (continued)

Question	Answer	Hint
8(c)	pH = 8 – 12	pH of salts are relatively straightforward: Strong acid + Strong alkali = Neutral salt Strong acid + Weak alkali = Acidic salt Weak acid + Strong alkali = Alkaline salt Weak acid + Weak alkali = Neutral salt Sodium hydroxide is a strong alkali and citric acid is a weak acid so the pH of the salt will be alkaline.
8(d)	Propanoic acid	The ester in this question is drawn with the section from the acid drawn first. Always look for the C=O as this indicates the acid part of the ester molecule.
8(e)	Concentrated sulphuric acid	It is essential to learn all the PPAs in detail. There will be at least two PPA based questions in each exam. The PPA covered here is from Unit 1 – 'Preparation of an Ester'
8(f)	25·35 g	Top Tip for percentage yield questions 1. Remember that all esterification reactions are a 1:1 ratio of alcohol to ester. So there is no need to balance the equation. Worked answer. <u>Step 1</u> Work out the moles of alcohol using Moles = mass ÷ formula mass. $(22·2 \div 74 = 0·3$ moles) <u>Step 2</u> The ratio is 1:1 so 0·3 moles of alcohol would produce 0·3 moles of ester. <u>Step 3</u> Calculate the theoretical mass of ester produced using Mass = moles × formula mass $(0·3 \times 130 = 39$ g) **(1 mark)** <u>Step 4</u> Calculate the percentage actual mass of ester produced $(39$ g × 65% = 25·35 g) **(1 mark)**
9(a)	An amino acid that is required in our diets as it cannot be produced by the body.	
9(b)	11 hydrogen atoms	This is an easy question to be caught out by. Remember that the phenyl group also has hydrogen atoms attached.

SECTION B (continued)

Question	Answer	Hint
9(c)		Remove an OH from the acid group of phenylalanine and an H from the amine group of the alanine and combine the two.

SECTION B (continued)

Question	Answer	Hint
9(d)	Peptide link	
10(a)	$2HCl + CaCO_3 \rightarrow CaCl_2 + H_2O + CO_2$	Should be fairly straightforward.
10(b)		A tricky question that requires some thought. Use $M = C \times V$ to calculate the number of moles of acid used in each experiment. This will show you that experiment one and two use the same number of moles of acid and would therefore produce the same volume of gas. However experiment two would produce the gas more quickly.
10(c)	1·1 g	Top Tip for questions involving $n = C \times V$ Remember that the volume must be in litres. Change cm^3 into litres by dividing by 1000. Two moles of acid produces one mole of carbon dioxide. Step 1 Establish the molar ratio. Step 2 Calculate the moles of acid using $n = C \times V$ ($1 \times 0.05 = 0.05$ moles) Step 3 Use the ratio to establish the moles of carbon dioxide produced. ($0.05 \div 2 = 0.025$ moles) Step 4 Calculate the mass of carbon dioxide using mass = moles × formula mass ($0.025 \times 44 = 1.1$ g)
11(a)	$Zn \rightarrow Zn^{2+} + 2e^-$	Ion-electron equations can be difficult but remember that most of them are in your data book!

SECTION B (continued)

Question	Answer	Hint
11(b)	CO₂ gas / HCl(aq) / CaCO₃(s) / Stage 1 — lime water turns milky as CO₂ is absorbed / Stage 2	Always make sure the diagram is clear, large and labelled.
11(c)	Incomplete combustion of the fuel	This is a simple standard grade question.
12(a)	**Compound X** (1 mark) **Compound Y** (1 mark)	There are two possible answers for compound Y.

SECTION B (continued)

Question	Answer	Hint
12(b)	Y $$\begin{array}{ccc} & H & Cl \\ & | & | \\ H- & C-C & -H \\ & | & | \\ & Cl & H \end{array}$$	The structure should be an isomer of the structure that you gave as your answer to question 12(a).
12(c)	Hydration	It is very important to learn all the types of reaction. Write them all down in a table and give an example of each.
12(d)	$\Delta H = 226$ kJ mol^{-1}	Top Tip for Hess's Law questions 1. Remember to balance the equations. 2. Remember to reverse the enthalpy sign when reversing an equation. 3. Remember to multiple the enthalpy value when multiplying an equation. Working You will need the balanced equations for the combustion of carbon, hydrogen and ethyne. (1 mark) **Equation 1** $C(g) + O_2(g) \longrightarrow CO_2(g)$ $\Delta H = -394$ kJ mol^{-1} **Equation 2** $H_2(g) + \frac{1}{2}O_2(g) \longrightarrow H_2O(g)$ $\Delta H = -286$ kJ mol^{-1} **Equation 3** $C_2H_2(g) + 2\frac{1}{2}O_2(g) \longrightarrow 2CO_2(g) + H_2O(l)$ $\Delta H = -1300$ kJ mol^{-1} To match the target equation, equation 1 has to be multiplied by two and equation 3 must be reversed. The enthalpies can then be combined. (1 mark)
13(a)	Substance A is Covalent molecular Substance B is Covalent network Substance C is Ionic Substance D is Metallic	It is essential to know all the types of bonds and structures as well as the properties associated with them.

SECTION B (continued)

Question	Answer	Hint
13(b)	Methanol is held together by hydrogen bonds due to the polarised –O-H group (1 mark). Methane molecules are held together by van der waals' forces which are much weaker than hydrogen bonds (1 mark).	In this type of question you must give as much information as possible in your answer.
14(a)	Hydrogenation is the addition of hydrogen across a double bond.	Refer back to question 12(c). If you produced and studied a table of reactions then this should have been an easy mark.
14(b)	Esters	Fats and oils can be classed as esters due to their functional group. Remember that Ester is a COO!
14(c)	12·6g	Step 1 Calculate the moles of sodium using Moles = mass ÷ formula $(0·1 ÷ 23 = 0·0043$ moles) Step 2 Establish the moles of sodium chloride. 0·0043 moles of sodium is equal to 0·0043 moles of sodium chloride because of the formula NaCl. (1 mark) Step 3 Calculate the mass of NaCl in 10 g of margarine using mass = moles × formula mass $(0·0043 × 58·5 = 0·25$ g) Step 4 Multiply by 50 to calculate the mass of salt in 500 g (1 mark)
15(a)	To provide the hydrogen ions	This question is poorly done in exams. The clue is in the equation.
15(b)	$0·158$ mol l^{-1}	Top tip for redox titration calculations Remember never to include the first (rough) titration in your average. Step 1 Establish the molar ratio. Two moles of MnO_4 reacts with 5 moles of H_2O_2. (It is easier to use the ratio of 1:2·5) Step 2 Calculate the moles of permanganate ions using $N = C × V$ $(0·1 × 0·0158 = 1·58 × 10^{-3})$ (1 mark) Step 3 Use the ratio to establish the moles of H_2O_2 used $(1·58 × 10^{-3} × 2·5 = 3·95 × 10^{-3}$ moles of $H_2O_2)$ (1 mark) Step 4 Calculate the concentration of the permanganate solution using $C = N ÷ V$. $(3·95 × 10^{-3} ÷ 0·025 = 0·158$ mol l^{-1}). (1 mark)

SECTION A

Question	Answer	Hint
1	D	Graphite is the only form of carbon that conducts electricity.
2	A	Electrons travel from the most reactive metal to the least reactive metal.
3	A	The negative terminal of a battery supplies the iron with electrons slowing down the oxidation process.
4	C	In both experiments the acid is in excess and therefore all the zinc will react to producing the same volume of gas.
5	D	This is a straightforward question. Sometimes they are not as difficult as they first appear!
6	C	The enthalpy change for the forward reaction is the difference in potential energy between the reactants and the products.
7	A	An increase in temperature increases the kinetic energy of the particles which results in more successful collisions. Only a catalyst can affect the activation energy of a reaction.
8	D	When water is melted the intermolecular bonds are broken and in the case of water, the intermolecular bonds broken are hydrogen bonds.
9	D	Electronegativity increases going from left to right across the periodic table with fluorine being the most electronegative. See Data Booklet page 10 for electronegativity values.
10	C	First ionisation is the energy required to remove **one mole of electrons** from one mole of atoms in the **gaseous state**.
11	D	Alkalis are produced from soluble metal oxides (bases).
12	A	This is easier than it first appears. The greatest number of moles gives the greatest number of molecules. All you have to do is work out moles using Moles = mass $\div$ formula mass.
13	B	Fullerenes (also know as buckminsterfullerene or buckie balls) are molecular compounds with approximately 60 carbon atoms per molecule.
14	B	Write out the balanced equation for the combustion of methane and you will see that 1 mole of methane reacts with 2 moles of oxygen. This leaves 1 mole of oxygen unreacted. The reaction also produces 1 mole of carbon dioxide so overall 2 moles of gas are present in the resulting gas mixture (1 mole of O_2 and 1 mole of CO_2). Water is not included because it is a liquid at room temperature.
15	D	This is a difficult question. One mole of aluminium sulphate has 5 moles of ions because of its formula $Al_2(SO_4)_3$. This means that 0.2 moles has ($0.2 \times 5 =$) 1 mole of ions.
16	C	Nitrogen dioxide is produced by the sparking of air in a petrol engine.
17	D	Learn all the types of reaction because they appear more than once in most exams.
18	D	The advantages and disadvantages of alternative fuels is very topical and will therefore appear in most exam papers.
19	B	To do this question correctly it is **NOT** essential to draw out the full structural formulas of the compounds. Simply count the number of carbon atoms present in each molecule. 3,3-dimethylbutan-1-ol has 6 carbon atoms. Only molecule B has 6 carbon atoms.
20	A	Learn all the types of reaction because they appear more than once in most exams.

SECTION A (continued)

Question	Answer	Hint
21	B	Don't be put off by the complicated structure. The molecule may appear very complicated but all you have to do is identify the two functional groups.
22	C	This is a difficult question and tests your knowledge of carbohydrates from Standard Grade.
23	B	Ketones are not readily oxidised.
24	B	Ozone absorbs UV radiation.
25	C	A tertiary alcohol is one in which the hydroxyl group is attached to a carbon atom that has no hydrogen atoms attached.
26	B	Esters are formed in the reaction between alcohols (hydroxyl group) and carboxylic acids (carboxyl group).
27	D	All plastics are carbon based compounds and will therefore give off carbon dioxide when burned, also some carbon monoxide if the oxygen is limited.
28	B	Enzymes are classed as proteins and when they are denatured they change shape which reduces their effectiveness. This can be explained using 'Lock and Key'.
29	D	Glycerol is a triol because it has three hydroxyl functional groups.
30	C	Learn all the types of reaction because they appear more than once in most exams. Remember that dehydration is the removal of water; dehydrogenation is the removal of hydrogen.
31	C	The blast furnace, Haber process and the Ostwald process require a lot of heat to operate efficiently so it is more economical for these to be continuous processes.
32	B	This question is very difficult. Make equation C your target equation and it becomes a simple Hess's Law question.
33	D	It is essential to write out the equation $NH_3 + H_2O \longrightarrow NH_4^+ + OH^-$ to answer this question correctly. The equilibrium will shift to the left to compensate which reduces the OH^- ion concentration. NH_4^+ does not decrease because ammonium is being added.
34	C	High pressure favours the end with the lowest gas volume and exothermic reactions favour a low temperature.
35	B	A pH change of 1 is a tenfold change in the H^+ ion concentration. So a pH change of 4 to 6 is a decrease in concentration by a factor of 100.
36	C	Manipulate the ionic product of water equation to calculate the pH like so: $[H^+] \times [OH^-] = 10^{-14}$ therefore $[H^+] = 10^{-14} \div [OH^-]$ ($10^{-14} \div 0.01 = 1^{-12}$)
37	D	Metal carbonates react with acids to produce carbon dioxide.
38	C	This is a difficult question. The equations must be multiplied and combined to cancel out the electrons like so: $3Cu(s) \rightarrow 3Cu^{2+}(aq) + 6e^-$ and $2NO_3^-(aq) + 8H^+(aq) + 6e^- \rightarrow 2NO(g) + 2H_2O(l)$. This results in a nitrate to copper molar ratio of 2:3. So 1 mole of nitrate oxidises 1·5 moles of copper.
39	D	An alpha particle (^4_2He) has a mass of four so repeatedly take four away from 227 until you reach 211.
40	B	Half-life is the time taken for a samples count/mass to fall by half.

SECTION B

Question	Answer	Hint
1(a)	Helium – Monatomic gas Graphite – Covalent network solid Oxygen – covalent molecular gas Potassium – metallic solid (½ mark for each correct answer)	Remember that Helium is a noble gas and as such doesn't form bonds.
1(b)	The electrons are not free to move.	Electrons must be free to move for an element to conduct electricity.
2(a)	Glycerol or propane-1,2,3-triol	By now this should be a fairly straight forward question.
2(b)	Fat molecules can pack closer together which allows the formation of intermolecular bonds.	Look up a diagrammatic representation of fats and oils as this makes it very clear why the boiling points are so different.
2(c)	Esters	Fats and oils all have the COO functional group.
3(a)	A catalyst	Catalysts lower the activation energy of a reaction but have no effect on the overall enthalpy change.
3(b)	The activated complex	The activated complex only exists for a fraction of a second and is an intermediary stage formed during the reaction.
3(c)	Exothermic	The reaction is classed as exothermic because the products have less energy than the reactants, so energy must have been given out in the formation of the products.
4(a)	X = 22 s, Y = 0·3 s^{-1}	It is essential to learn all the PPAs in detail. There will be at least two PPA based questions in each exam. The PPA covered here is from Unit 1 – 'Effect of Concentration on Reaction Rate'
4(b)	Colourless to blue/black	
4(c)	Correct scale (1 Mark) Axis labelled correctly (1 Mark) Points plotted correctly (1 Mark) *(graph: Rate/s^{-1} vs Volume of KI/cm^3, axis values 0·01 0·02 0·03 0·04 0·05 and 0 5 10 15 20 25)*	To gain all the marks in graph questions make sure you do the following: 1. Make sure the graph is as big as the graph paper allows. 2. Always draw a line of 'best fit' using a pencil 3. Label both axis correctly and clearly including the units 4. Make sure the points are correctly plotted
4(d)	To ensure that the concentration of all the other reactants remains constant.	This question should be straight forward if you have studied and learned all your PPA reports.

SECTION B (continued)

Question	Answer	Hint
5 (a)	Incomplete combustion or loss of ethanol due to evaporation	It is essential to learn all the PPAs in detail. There will be at least two PPA based questions in each exam. The PPA covered here is from Unit 1 – 'Enthalpy of Combustion'
5 (b)	-1345 kJ mol^{-1} Partial marks for: Using $\Delta H = cm\Delta T$ to calculate the enthalpy change for the reaction. (1 Mark) Calculating the moles of ethanol used. (1 Mark) Calculating the enthalpy of combustion of 1 mole of ethanol. (1 Mark)	Top Tips for questions involving $\Delta H = cm\Delta T$ 1. Remember that 100 cm^3 of water is 0·1 kg 2. Remember that all combustion reactions are exothermic and ΔH should therefore have a negative charge Worked Answer Step 1 Calculate the enthalpy change using $\Delta H = cm\Delta T$ ($4\cdot18 \times 0\cdot4 \times 16\cdot1$ $= -26\cdot9$ kJ mol^{-1}) Step 2 Calculate the moles of ethanol used using Moles = mass ÷ formula ($0\cdot92 \div 46 = 0\cdot02$ moles) Step 3 Use this to calculate the enthalpy for 1 mole using 0·02 moles produces $-26\cdot9$ kJ mol^{-1} then 1 mole produces ($1 \div 0\cdot02 \times -26\cdot9 = -1345$ kJ mol^{-1})
6 (a)	Displacement	Displacement reactions involve both oxidation and reduction reactions.
6 (b)	The UF$_4$ would react with the oxygen.	Reactions are often carried out in atmospheres of inert gases such as the noble gases and nitrogen. This avoids unwanted reactions.
6 (c)	The half-life would be the same.	Half-life is independent of temperature, pressure, concentration and chemical state of the isotope.
6 (d)	Covalent bonding (molecular)	The low melting point of 64.8°C suggests that the bonding is covalent.
7 (a)	Mg + 2HCl ⟶ MgCl$_2$ + H$_2$	Markers assess these questions using 'Will the experiment work as drawn?' Ensure that the diagram is clearly labelled and not too small.
7 (b)		

SECTION B (continued)

Question	Answer	Hint
7(c)	24·4 litres	Top Tips for questions involving molar volume. 1. Treat them the same as any other calculation. (They are not as hard as you think!) 2. Molar volume is in litres so all volumes used in the calculation must also be in litres. Worked Answer Step 1 Establish the molar ratio from the balanced equation (1 mole of Mg produces 1 mole of H_2) Step 2 Calculate the moles of Magnesium used using Moles = mass ÷ formula $(0·2 ÷ 24·3 = 0·0082$ moles) (1 mark) Step 3 Use the molar ratio to establish the number of moles of hydrogen produced. (0·0082 moles) Step 4 Using the fact that 0·0082 moles occupies a volume of 0·22 litres, calculate the molar volume $(1 ÷ 0·0082 × 0·2 = 24·4$ litres) (1 mark)
7(d)	To ensure that all the magnesium had reacted	For the calculation to be correct all of the 0·2 g of magnesium must react.
8(a)	Body centred cubic	This question may seem very difficult (even scary!) at first but it is quite straight forward once you read it carefully. To calculate the radius ratio use the formula given and the information on page 16 of the data book. $(65 ÷ 136 = 0·48)$
8(b)	Potassium oxide doesn't fit the formula XY	Potassium oxide has the formula K_2O and therefore doesn't fit the formula XY

SECTION B (continued)

Question	Answer	Hint
9(a)	Negative electrode	'Opposites attract' so the positively charged silver ion would be formed at the negative electrode
9(b)	0·0595 litres or 59·5 cm^3	This is a very difficult calculation! Top Tips 1. You would expect this question to involve $Q = I \times T$ but there is no information about time and current used so $Q = I \times T$ cannot be used here. Worked Answer Step 1 Establish the quantity of electricity passed to produce 1 mole of silver (96 500 C because silver has a 1 positive charge) Step 2 Calculate the moles of silver produced using Moles = mass ÷ formula (0·54 ÷ 107·9 = 0·005 moles) (1 mark) Step 3 Using the fact that 1 mole of silver is produced by 96 500 C, calculate the quantity of electricity passed (0·005 × 96 500 C = 482·9 C) Step 4 Establish the quantity of electricity passed to produce 1 mole of hydrogen (19 300 C = 96 500 × 2) because hydrogen has a 1 positive charge but is diatomic) Step 5 Using the fact that 1 mole of hydrogen is produced by 19 300 C calculate the number of moles of hydrogen produced by 482·9 C (482·9 ÷ 193 000 C = 0·0025 moles) (1 mark) Step 6 Calculate the volume of hydrogen occupied by 0·0025 moles (0·0025 × 23·8 = 0·05951) (1 mark)
10	Methanoic acid dissociates only partially in water as shown by the equation HCOOH ⇌ HCOO$^-$ + H$^+$ Hydrochloric acid dissociates completely in water as shown by the equation HCl → H$^+$ + Cl$^-$ (1 mark) This results in a high pH for methanoic acid as it has a lower concentration of H$^+$ ions. (1 mark) Methanoic acid has a lower conductivity because it has a lower concentration of ions. (1 mark)	This question tests all your knowledge of strong and weak acids.
11(a)	Record the volume of oxygen gas collected with time and repeat the experiment at a different temperature but keeping all other variables constant.	Question 11(a) is based on the PPA 'Factors Effecting Enzyme Activity'
11(b)	37°C	Most enzymes work best at body temperature

SECTION B (continued)

Question	Answer	Hint
11(c)	Reactant 1 Reactant 2 Product	
12(a)		Always try this type of question on spare paper and check that each element is forming the correct number of bonds.
12(b)	Addition polymerisation	The molecule is unsaturated which suggests addition polymeristaion.
12(c)	$2CH_2CHCN + H_2O + H^+ + 2e^- \rightarrow (CH_2CH_2CN)_2 + OH^-$	A tricky one. Follow these steps to get the answer 1. Balance the equation 2. Balance the oxygen atoms by adding water to the left hand side. 3. Balance the hydrogen atoms by adding H^+ to the left hand side. 4. Balance the charge by adding $2e^-$ to the left hand side.
13(a)	Test with bromine solution	Benzene does not decolourise bromine solution however Kekulé's structure suggests that it would decolourise bromine because of the double bonds.
13(b)		
13(c)	It increases the octane number	Benzene is added to unleaded petrol as a substitute for lead compounds to improve the efficiency of burning.

SECTION B (continued)

Question	Answer	Hint
13(d)	$\Delta H = 46$ kJ mol^{-1}	Top Tip for Hess's Law questions 1. Remember to balance the equations. 2. Remember to reverse the enthalpy sign when reversing an equation. 3. Remember to multiply the enthalpy value when multiplying an equation. Working You will need the balanced equations for the combustion of carbon, hydrogen and benzene. (1 mark) **Equation 1** $C(g) + O_2(g) \rightarrow CO_2(g)$ $\Delta H = -394$ kJ mol^{-1} **Equation 2** $H_2(g) + \tfrac{1}{2}O_2(g) \rightarrow H_2O(g)$ $\Delta H = -286$ kJ mol^{-1} **Equation 3** $C_6H_6(l) + 7\tfrac{1}{2}O_2(g) \rightarrow 6CO_2(g) + 3H_2O(l)$ $\Delta H = -3268$ kJ mol^{-1} To match the target equation, equation 1 has to be multiplied by six and equation 2 has to be multiplied by three. Equation three must be reversed. (1 mark) The enthalpies can then be combined. (1 mark)
14(a)	Molecule A is but-2-ene Molecule B is 2-chlorobutane Molecule C is butan-2-ol	Draw out the full structural formulas of the molecules to make this question easier.
14(b)	Hydrogen chloride	
14(c)	**(1 mark)** **Butanone (1 mark)**	The position of the carbonyl group does **not** have to be numbered because it can only be on the second carbon in butanone.
14(d)	Tollens reagent	
14(e)	Acidified potassium dichromate, hot copper(II) oxide.	Any suitable oxidising agent.